Fractional Order Systems

Fractional Order Systems

Special Issue Editor
Ivo Petráš

MDPI • Basel • Beijing • Wuhan • Barcelona • Belgrade

MDPI

Special Issue Editor
Ivo Petráš
Technical University of Košice
Slovakia

Editorial Office
MDPI
St. Alban-Anlage 66
4052 Basel, Switzerland

This is a reprint of articles from the Special Issue published online in the open access journal *Mathematics* (ISSN 2227-7390) in 2019 (available at: https://www.mdpi.com/journal/mathematics/special_issues/Fractional_Order_Systems)

For citation purposes, cite each article independently as indicated on the article page online and as indicated below:

LastName, A.A.; LastName, B.B.; LastName, C.C. Article Title. *Journal Name* **Year**, *Article Number, Page Range.*

ISBN 978-3-03921-608-6 (Pbk)
ISBN 978-3-03921-609-3 (PDF)

Contents

About the Special Issue Editor

Ivo Petráš (Professor, PhD., DrSc.) received his MSc. (1997), PhD. (2000), and DrSc. (2013) degree in process control at the Technical University of Kosice and the Slovak University of Technology in Bratislava, Slovak Republic, respectively. He works at the Institute of Control and Informatization of Production Processes, Faculty of BERG, Technical University of Kosice as a Professor. His research interests include control systems, automation and applied mathematics. He is a member of IEEE, IFAC, and SSAKI. He has published five books, five book chapters, over 70 journal papers, and over 90 refereed conference papers.

Preface to "Fractional Order Systems"

Fractional calculus deals with the consideration of integrals and derivatives of arbitrary order (constant, variable, and distributed) as well as also with integro-differential equations, so called, fractional differential equations. Historically, fractional calculus has been recognized since the inception of regular calculus, with the first written reference dated in September 1695 in a letter from Leibniz to L'Hospital. Nowadays, fractional calculus has a wide area of applications in areas such as physics, chemistry, bioengineering, chaos theory, control systems engineering, and many others. In all those applications, we deal with fractional order systems in general. Moreover, fractional calculus plays an important role even in complex systems and therefore allows us to develop better descriptions of real-world phenomena. On that basis, fractional order systems are ubiquitous, as the whole real world around us is fractional. Due to this reason, it is urgent to consider almost all systems as fractional order systems.

This Special Issue is focused on the theory and multidisciplinary applications of fractional order systems in science and engineering. It consists of the following collection of papers:

Fractional Calculus as a Simple Tool for Modeling and Analysis of Long Memory Process in Industry by Ivo Petráš and Ján Terpák.

Back to Basics: Meaning of the Parameters of Fractional Order PID Controllers by Inés Tejado, Blas M. Vinagre, José Emilio Traver, Javier Prieto-Arranz and Cristina Nuevo-Gallardo.

Audio Signal Processing Using Fractional Linear Prediction by Tomas Skovranek and Vladimir Despotovic.

Time-Fractional Diffusion-Wave Equation with Mass Absorption in a Sphere under Harmonic Impact by Bohdan Datsko, Igor Podlubny and Yuriy Povstenko.

Fractional Order Complexity Model of the Diffusion Signal Decay in MRI by Richard L. Magin, Hamid Karani, Shuhong Wang and Yingjie Liang.

Adaptive Pinning Synchronization of Fractional Complex Networks with Impulses and Reaction–Diffusion Terms by Xudong Hai, Guojian Ren, Yongguang Yu and Conghui Xu.

Optimal Randomness in Swarm-Based Search by Jiamin Wei, YangQuan Chen, Yongguang Yu and Yuquan Chen.

These papers reflect the latest research achievements in the field of fractional calculus and its applications in various areas.

Ivo Petráš
Special Issue Editor

mathematics

MDPI

Article

Fractional Calculus as a Simple Tool for Modeling and Analysis of Long Memory Process in Industry

Ivo Petráš and Ján Terpák

Faculty of BERG, Technical University of Košice, Němcovej 3, 04200 Košice, Slovakia; jan.terpak@tuke.sk
* Correspondence: ivo.petras@tuke.sk; Tel.: +421-55-602-5194

Received: 25 April 2019; Accepted: 3 June 2019; Published: 4 June 2019

Abstract: This paper deals with the application of the fractional calculus as a tool for mathematical modeling and analysis of real processes, so called fractional-order processes. It is well-known that most real industrial processes are fractional-order ones. The main purpose of the article is to demonstrate a simple and effective method for the treatment of the output of fractional processes in the form of time series. The proposed method is based on fractional-order differentiation/integration using the Grünwald–Letnikov definition of the fractional-order operators. With this simple approach, we observe important properties in the time series and make decisions in real process control. Finally, an illustrative example for a real data set from a steelmaking process is presented.

Keywords: fractional calculus; fractional-order system; long memory; time series; Hurst exponent

---

1. Introduction

In this paper, we discuss how the fractional calculus is used in modeling and analysis of fractional-order processes (e.g., a real industrial process). Fractional-order processes are characterized by long memory, local memory or heavy-tailed distributions. These properties cannot be neglected in time series analysis. The fractional calculus provides a tool for both long memory and local memory process analysis.

Long memory processes are known to play an important role in many areas of science and technology. The idea of applying a fractional-order model in time series analysis is not new. In the last 20 years, a significant progress has been made in understanding the probabilistic foundations and statistical principles of such processes. For instance, the Hurst exponent was used as a measure of long-term memory of time series [1]. In [2] a fractionally differenced autoregressive-moving average process was used. Nowadays, fractional calculus, fractional-order systems, fractional-order processes and fractional signal processing techniques and their real world applications in various areas have been described in several works [3–8]. Long memory has been observed even in economics [9]. A very useful application of the fractional calculus in the time series analysis of heart rate variability was shown in [10]. Some new fractional (a.k.a. fractal) time series models and processes have been discussed in the tutorial review [11].

In this article, we propose a useful tool for applied researchers and practitioners who need to analyze data in which power laws, long memory, self-similar scaling or fractal properties are relevant. We combine the two modeling approaches: statistical analysis and the fractional calculus.

2. Preliminaries

2.1. Fractional Calculus

Fractional calculus has been known since the beginning of integer-order calculus, as it dates back to the 1695 correspondence between Leibniz and L'Hospital. It is a generalization of integration and differentiation to a non-integer order fundamental operator $_aD_b^\alpha$, where a and b are the bounds of the operation and α is an arbitrary order. The usual notation for the left-sided fractional-order integro-differential operator of a function $f(t)$ defined within the interval $t > 0$ is $_0D_t^\alpha f(t) \equiv d^\alpha f(t)/dt^\alpha \equiv f^{(\alpha)}(t)$, with $\alpha \in$ R. The basic theory was developed mainly in the 19th century, and engineering applications were realized mainly in the second half of the 20th century. There exist a lot of definitions of fractional-order operators (integrals and derivatives) for constant, variable, distributed and even complex order. However, in this paper, we consider only the Grünwald–Letnikov definition of constant order, which is equivalent to other definitions (Caputo, Riemann–Liouville) for a wide class of the functions, under certain conditions [5].

Definition 1. *The Grünwald–Letnikov definition of the fractional-order operator is given as [5]:*

$$_aD_t^\alpha f(t) = \lim_{h \to 0} \frac{1}{h^\alpha} \sum_{j=0}^{[\frac{t-a}{h}]} (-1)^j \binom{\alpha}{j} f(t - jh),$$ (1)

where [.] means the integer part.

The form (1) of the fractional operator definition is very helpful for finding a numerical approximation of fractional derivatives/integrals as well as the solution of fractional differential equations. For the definition of binomial coefficients, we may use the relation based on Euler's *Gamma* function, defined as follows:

$$\binom{\alpha}{j} = \frac{\Gamma(\alpha + 1)}{\Gamma(j + 1)\,\Gamma(\alpha - j + 1)},$$ (2)

where $\binom{\alpha}{0} = 1$.

Some other definitions of fractional-order operators, useful properties, special transforms and methods for analytical and numerical solutions of fractional differential equations can be found, for example, in [4–6,12].

2.2. Fractional-Order Processes

It is well-known that a fractional-order process can be considered as the output of a fractional-order system, and has long memory properties. The essence of fractional-order processes is a power-law, which demonstrates the long memory itself. Generally, there is a large number of real processes, where the fractional calculus can be applied [8]. Westerlund proposed a description of a "universal process" by the following equation [13]:

$$y(\tau) = k\frac{d^\alpha x(\tau)}{d\tau^\alpha} \equiv k\,_0D_\tau^\alpha x(\tau),$$ (3)

where the input, $x(\tau)$, is an intensive signal, the output, $y(\tau)$, is extensive, and parameters, k and α, are the process constants. For example, for electrical processes, the electrical current is an extensive signal, and the electrical voltage is an intensive signal. Similarly, for mechanical processes, it is force and position; for heat processes, it is heat flux and temperature, and so on [14]. Thus, the fractional-order processes are widely found in science, technology and engineering systems [6].

Moreover, the use of fractional-order derivatives for the description and analysis of real processes also requires their geometrical interpretation [3,15–18]. For example, let us consider the function $f(\tau) = \tau^2$ as an intense signal. The line passing through the point $P(\tau_0, f(\tau_0))$ is given by the equation

$$f(\tau) - f(\tau_0) = f^{(\alpha)}(\tau_0)(\tau - \tau_0),$$ (4)

where $f^{(\alpha)}(\tau_0)$ is the slope of the line. Figure 1 shows the plot of a function with lines formed with fractional order derivatives at the point $P(1,1)$.

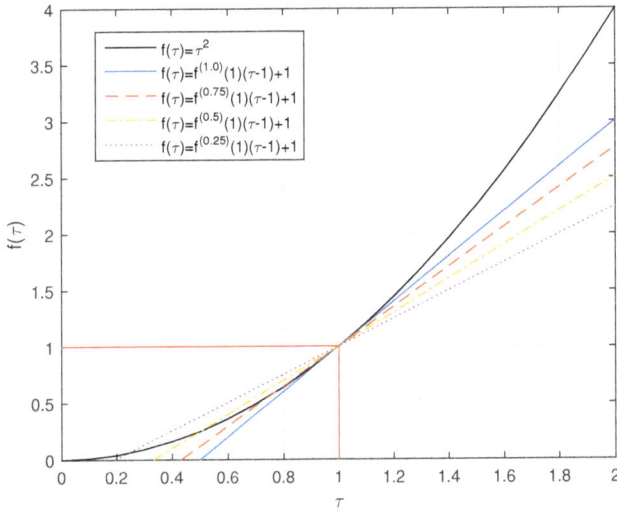

Figure 1. Plot of the function $f(\tau) = \tau^2$ with lines formed with fractional-order derivatives at the point $P(1,1)$.

In the case of the first-order derivative, the slope of the line is equal to the tangent line. The tangent angle of the line slope at a given point in the function depends on the value of the fractional-order derivative at the considered point.

Definition 2. *According to the Grünwald–Letnikov definition (1), the tangent of the angle θ can be expressed as follows:*

$$\tan \theta = f^{(\alpha)}(\tau) \approx \frac{\sum_{j=0}^{\infty} b_j f(\tau - jh)}{h^\alpha} \tag{5}$$

or

$$\tan \theta = f^{(\alpha)}(\tau) \approx \frac{f(\tau) + \sum_{j=1}^{\infty} b_j f(\tau - jh)}{h^\alpha}, \quad \text{with} \quad b_0 = 1, \tag{6}$$

where binomial coefficients b_j represent the weight of the effect of the function history, respectively, the sum $\sum_{j=1}^{\infty} b_j f(\tau - jh)$ partially accumulate the history of the function. For the binomial coefficients b_j, an expression (8) can be used.

2.3. Fractional Time Series

Fractional time series are characterized by the Hurst exponent (parameter). The Hurst exponent H was originally developed by Harold Edwin Hurst in hydrology for determining optimal dam sizing for the Nile river. It is directly related to fractal dimension D, such that $D = 2 - H$. The Hurst exponent is defined as follows [19–21]:

Definition 3. *Let us consider d as a duration, which is a period of time that includes several points in the time series over the range R as a difference between the largest and smallest deviation encountered during a duration d, then we can write:*

$$R \propto d^H, \quad (0 < H < 1),$$

where H is the Hurst exponent. It means that the higher the Hurst exponent is, the smoother the curve is.

The Hurst exponent is used as a measure of long memory in a time series. It takes on values from 0 to 1. A value of 0.5 indicates the lack of long-range dependence, and the time series has no statistical dependence. Absolute values of the autocorrelation functions decay exponentially to zero. A parameter H less than 0.5 corresponds to anti-persistency. When H exceeds one half and is closer to 1, it indicates greater degree of persistence or long-range dependence. For the cases $0.5 < H < 1$ and $0 < H < 0.5$, power-law decay is typically observed. Time series that are Gaussian may be analyzed efficiently, but those which exhibit anti-persistence (negatively correlated process) or persistence (positively correlated process) resist simple analysis. By using the operation of fractional integration to an anti-persistent time series (or fractional differentiation to a persistent time series), Gaussian behavior can be observed.

3. Proposed Method and Analysis Tool

As was mentioned in the previous section, there exists a technique for correcting the data to remove the anti-persistence or persistence by applying fractional calculus of order $1/2 - H$. This yields data that obey Gaussian statistics and therefore the time series can be processed and analyzed. Thus, if a white Gaussian process is fractional differ-integrated with order $-\alpha$, then the acquired time series has the Hurst parameter equal to $\alpha + 1/2$. Fractional integration and differentiation are significant novelties in analysis because of the improved accuracy of estimates and properties of the series. If a series is not described properly, then further analyses will not be accurate. A review of methods for the estimation of the Hurst parameter, which are helpful as simple diagnostic tools for time series, was given in [6]. The R/S method is one of the most well-known estimators. A useful Matlab function for the exponent estimation is available [22].

Moreover, for numerical computation of the fractional-order operation, the following formula derived from the Grünwald–Letnikov definition (1), at the points kT ($k = 1, 2, \dots$), can be used [4]:

$$_aD^\alpha_{t_k} f(t) \approx T^{-\alpha} \sum_{j=0}^{k} b_j f(t_k - j),$$

(7)

where $t_k = kT$, T is the time step of calculation (sampling period) and the binomial coefficients b_j can be calculated according to recurrence relation:

$$b_0 = 1, \qquad b_j = \left(1 - \frac{1 + \alpha}{j}\right) b_{j-1}.$$

(8)

A Matlab function based on a discrete form of the relations (7) and (8) has been provided [23].

By applying both aforementioned techniques, we obtain a powerful tool for analysis of real (fractional-order) processes, especially in the case, when we do not have enough a priori information about the process due to a lack of measured information, etc.

4. Example: Modeling and Analysis of Industrial Process

4.1. Process Description

In this example, the data of the steelmaking process in a basic oxygen furnace located at U.S. Steel Košice, Ltd., Slovak Republic, are presented. The data of 240 melts were collected during the year 2018. The basic oxygen furnace is a pear shaped vessel, where pig iron from the blast furnace and ferrous scrap is refined into steel by blowing high-purity oxygen through the hot metal. This large vessel has a capacity of up to 400 tons of melt at high temperatures of 1650 to 1700 °C. It is a highly complex thermochemical process with high energy consumption. The heat energy is obtained by burning mainly carbon and silicon that are in the inputs. The basic material inputs to the basic oxygen furnace process are pig iron, ferrous scrap, slag additives, and blown oxygen. The outputs of the process consist of steel, slag, and waste gas [24,25].

Very important information about the process and at the end of the process is the concentration of carbon monoxide and carbon dioxide in the waste gas (see Figure 2 for one selected melt). At the start of the process, the concentration is zero, and the concentration then increases to maximum values, and at the end decreases to zero again. Similarly, the change of the total concentration of carbon oxides corresponds to the decarburization rate. In terms of process control, the decarburization rate below a certain threshold value means terminating the process in the basic oxygen furnace.

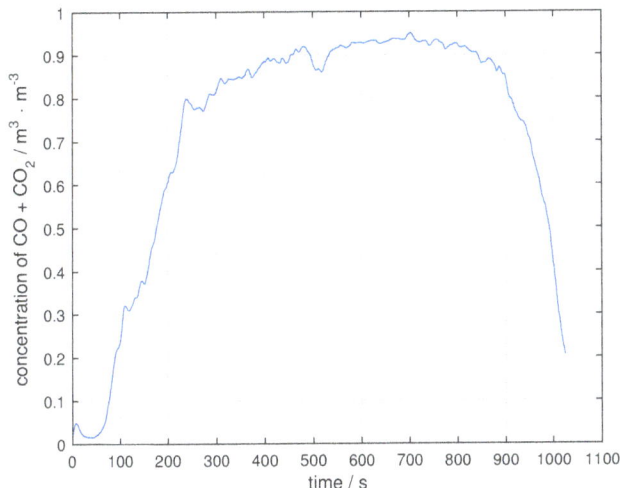

Figure 2. The change of CO and CO_2 concentration.

The decarburation rate can be given by

$$\frac{dm_C}{d\tau} = \frac{\phi_{wg}(CO_{wg} + CO_{2,wg})}{V_M},$$ (9)

where ϕ_{wg} is the measured waste gas flow, CO_{wg} and $CO_{2,wg}$ are measurement concentration of carbon monoxide and carbon dioxide in the waste gases, and V_M is the molar volume [26]. Calculation of the decarburization rate according to Equation (9) is only possible in the case where the waste gas flow is measured. If the waste gas flow is not measured, we need to find a method to determine the end of the process. Taking into account the information about of the fractional-order process, we applied the new proposed method.

4.2. Process Analysis

Signal depicted in Figure 2 is a long memory process and a typical long-range dependence time series, with the Hurst exponent $H = 0.98$, where R/S statistic and Hurst line is depicted as log-log plot in Figure 3. The slope of the line gives the Hurst exponent.

The signals, which exhibit fractional properties, should be investigated using the fractional calculus technique to obtain better analysis results. Using the relation (7) derived from the Grünwald–Letnikov definition (1), for $T = 1$ s, the change of the concentration of carbon oxides for various orders is shown in Figure 4. We used the same melt as is depicted in Figure 2. In the case of the first-order derivative, the values move in a narrow range around a value equal to zero, and, at the end of the process, there is a slight decrease. This is because the first-order derivative means roughly the difference of two consecutive values of the function. In the case of the derivative of order less than one, the values move above the zero and only fall below the limit value if the process ends. This follows from the memory effect, that is, that the value of the fractional-order derivative of the function is affected by all previous function values. The question is which order of derivative we should use.

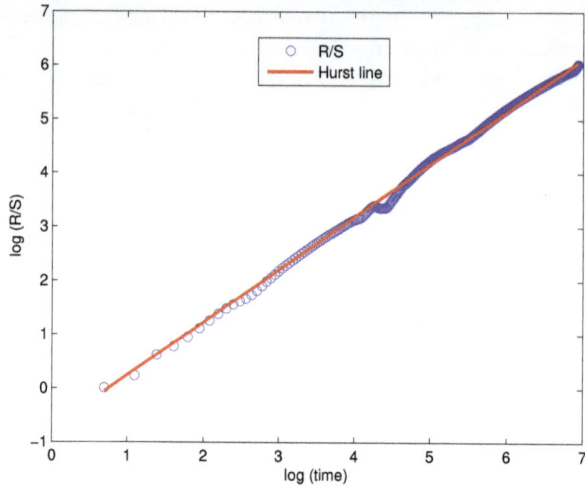

Figure 3. The slope of the Hurst line.

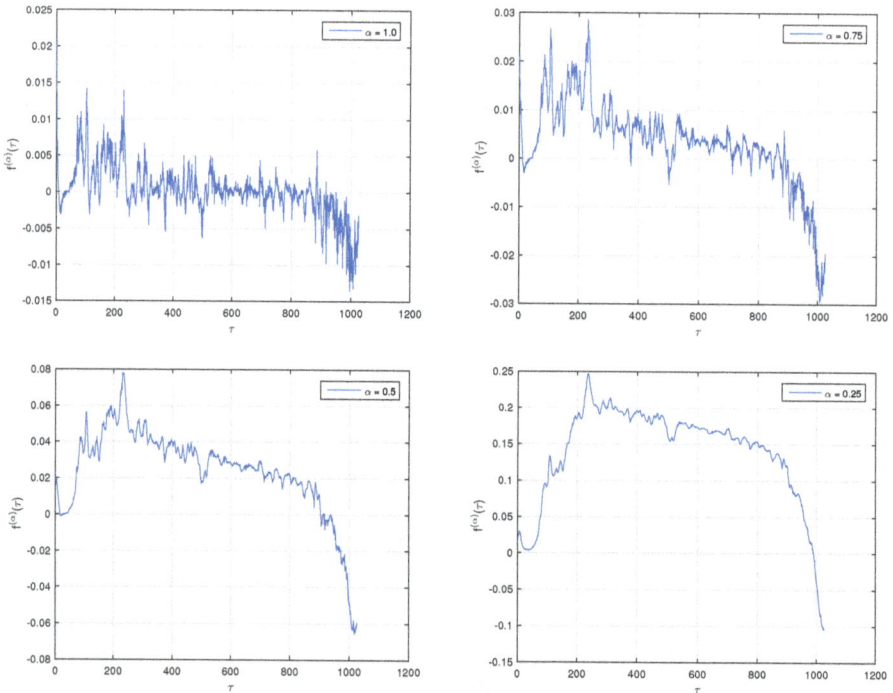

Figure 4. The change of CO and CO_2 concentration for various orders $\alpha \in (0, 1]$.

The use of half-order derivatives for the computation of heat flux, based on the known history of temperatures, was described in [5,12]. If we generalize this case, then it is possible to use half-order derivatives for the computation of the extensive signal behavior (heat flow, mass flow, electric current, and so on), based on the known behavior of the intensive signal (temperature, concentration, electrical potential, and so on).

In our case, the mass flow of carbon $(dm_C(\tau)/d\tau)$ is proportional to the concentration of carbon oxides $(x_{CO,CO_2}(\tau))$ as

$$\frac{dm_C(\tau)}{d\tau} \propto \frac{d^\alpha x_{CO,CO_2}(\tau)}{d\tau^\alpha}. \tag{10}$$

The data analysis of 240 melts was undertaken. Different orders of fractional differentiation within interval the $\alpha \in (0,1]$ were calculated for all concentrations of the considered melts, as it is depicted in Figure 4. Subsequently, the end point value or minimum value was determined. Table 1 shows and compares the values of the statistical indices for the orders α obtained from all 240 melts. Because of different values, a relative standard deviation in percent as a measure of variability was also calculated. The values indicate that the smallest relative standard deviation is between $\alpha = 0.4$ and $\alpha = 0.5$.

Table 1. Statistical values for the process end point.

α	Arithmetic Mean	Median	Range R	Standard Deviation S	Relative Standard Deviation
0.0	0.001552	0.001157	0.008391	0.001753	112.99
0.1	−0.110144	−0.110675	0.058425	0.012487	11.33
0.2	−0.135971	−0.135620	0.072973	0.013492	9.92
0.3	−0.124915	−0.125612	0.050907	0.010584	8.47
0.4	−0.102376	−0.101533	0.035958	0.007871	7.68
0.5	−0.079537	−0.078806	0.026901	0.006659	8.37
0.6	−0.060418	−0.060031	0.030330	0.006287	10.40
0.7	−0.046102	−0.044751	0.032936	0.006192	13.43
0.8	−0.035730	−0.034599	0.037782	0.006508	18.21
0.9	−0.028537	−0.027227	0.042185	0.006910	24.21
1.0	−0.023569	−0.021991	0.044271	0.007089	30.07

4.3. Simulation Results and Discussion

From the above results, it follows that, for the given device, we can determine the process endpoint using the fractional derivative of the concentration of carbon oxides. The value of the derivative order and the endpoint value should be determined for each basic oxygen furnace separately based on the analysis of the measured data during long period. It should be noted that, according to the fractional time series theory, the order of fractional derivative should correspond to the Hurst exponent of measured data in the time series. It is an effective tool that is useable for various long memory processes. For the melt depicted in Figure 2, the Hurst exponent $H = 0.98$ was obtained. For accurate determination of the end point in the process, we should analyze the fractionally differentiated time series with order $\alpha = 0.98 - 1/2 = 0.48$, which also corresponds to the results obtained by statistical analysis shown in Table 1.

In Figure 5, the change of CO and CO_2 concentration for order $\alpha = 0.48$ for one particular melt is depicted as it was shown in Figure 2. From the figure, we may determine the process end point which is extremely significant for the furnace operator. In the presented case, the process end point is approximately at 900 s. Since the steelmaking process is a process characterized by high energy consumption, the accurate finishing time of each melt is very important for minimizing production costs.

Moreover, the order of differentiation applied to the time series (measured data) moderates the curve shape (function) as it was demonstrated in Figure 1, as well as by the relations (5) or (6), where the history of the function is considered. In the case of a basic oxygen furnace, the function is the concentration of carbon oxides.

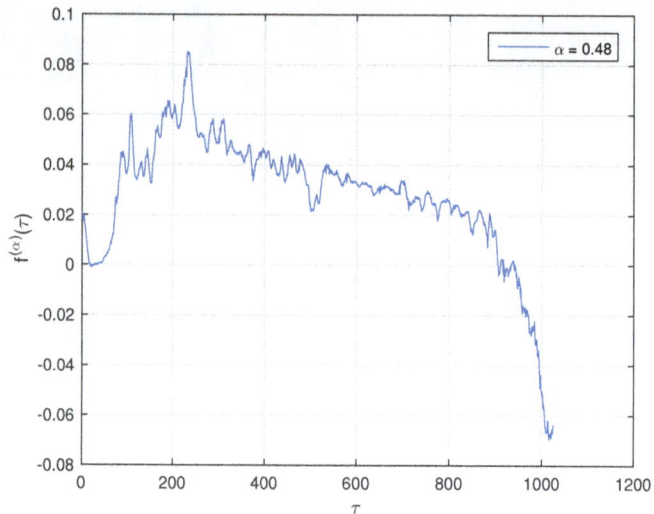

Figure 5. The change of CO and CO_2 concentration for order $\alpha = 0.48$.

5. Conclusions

In this paper, we described a simple tool for the modeling and analysis of long memory processes such as the steelmaking process. This approach is based on the Grünwald–Letnikov definition of fractional-order operator (derivative/integral). In this particular case, we used this definition because of data being presented in a time series, where the other definitions are not appropriate for such data processing since the data are given as a sequence, not a function. Obviously, for mathematical modeling where the process can be described by a function, it is more suitable to use Caputo's definition mainly because of better setting of initial conditions. However, another advantage for using the Grünwald–Letnikov definition is the possibility of using a short memory principle [5], especially in industry, where the hardware devices have limited memory and processor speed.

Combining two known mathematical tools, the fractional calculus technique and statistical methods, we obtain a simple but powerful tool to determine the process end point from measured data in the form of time series without direct measurement of the necessary exact value, in our case, the waste gas flow measurement. It is extremely important for industry to minimize costs as well as technical difficulties with making such a measurement.

We presented the techniques and tools for data processing with references for Matlab functions. Finally, an illustrative example of a real industrial process is given to demonstrate the effectiveness of obtained results.

Author Contributions: This work involved all coauthors. I.P. wrote the original draft and contributed to the software and the editing. J.T. analyzed the data, contributed to the visualization and investigated the results.

Funding: This research was funded by the Slovak Research and Development Agency under the contract No. APVV-14-0892, No. SK-AT-2017-0015, and by the Slovak Grant Agency for Science under Grant No. VEGA 1/0365/19 and No. VEGA 1/0273/17.

Conflicts of Interest: The authors declare no conflict of interest.

References

1.   Diebolt, C.; Guiraud, V. Long Memory Time Series and Fractional Integration. A Cliometric Contribution to French and German Economic and Social History. *Hist. Soc. Res.* **2000**, *25*, 4–22.
2.   Li, W.K.; McLeod, A.I. Fractional time series modelling. *Biometrika* **1986**, *73*, 217–221. [CrossRef]

3. Bandyopadhyay, B.; Kamal, S. *Stabilization and Control of Fractional Order Systems*; Springer: Cham, Switzerland, 2015.
4. Petráš, I. *Fractional-Order Nonlinear Systems: Modeling, Analysis and Simulation*; Springer: New York, NY, USA, 2011.
5. Podlubny, I. *Fractional Differential Equations*; Academic Press: San Diego, CA, USA, 1999.
6. Sheng, H.; Chen, Y.Q.; Qiu, T.S. *Fractional Processes and Fractional-Order Signal Processing*; Springer: London, UK, 2012.
7. Sun, H.G.; Zhang, Y.; Baleanu, D.; Chen, W.; Chen, Y.Q. A new collection of real world applications of fractional calculus in science and engineering. *Commun. Nonlinear Sci. Numer. Simulat.* **2018**, *64*, 213–231. [CrossRef]
8. West, B.J. *Fractional Calculus View of Complexity*; CRC Press: New York, NY, USA, 2016.
9. Tarasov, V.E.; Tarasova, V.V. Dynamic Keynesian Model of Economic Growth with Memory and Lag. *Mathematics* **2019**, *7*, 178. [CrossRef]
10. Garcia-Gonzalez, M.A.; Fernandez-Chimeno, M.; Capdevila, L.; Parrado, E.; Ramos-Castro, J. An application of fractional differintegration to heart rate variability time series. *Comput. Methods Programs Biomed.* **2013**, *111*, 33–40. [CrossRef] [PubMed]
11. Li, M. Fractal Time Series—A Tutorial Review. *Math. Probl. Eng.* **2010**, *2010*, 157264. [CrossRef]
12. Oldham, K.B.; Spanier, J. *The Fractional Calculus*; Academic Press: New York, NY, USA, 1974.
13. Westerlund, S. *Dead Matter Has Memory!*; Causal Consulting: Kalmar, Sweden, 2002.
14. Doebelin, E.O. *System Dynamics: Modeling and Response*; Merrill: Columbus, OH, USA, 1972.
15. Podlubny, I. Geometric and physical interpretation of fractional integration and fractional differentiation. *Fract. Calc. Appl. Anal.* **2002**, *4*, 357–366.
16. Nizami, S.T.; Khan, N.; Khan, F.H. A new approach to represent the geometric and physical interpretation of fractional order derivatives of polynomial function and its application in field of science. *Can. J. Comput. Math. Nat. Sci. Eng. Med.* **2010**, *1*, 1–8.
17. Machado, J.A.T. A probabilistic interpretation of the fractional-order differentiation. *Fract. Calc. Appl. Anal.* **2003**, *1*, 73–80.
18. Moshrefi-Torbati, M.; Hammond, J.K. Physical and geometrical interpretation of fractional operators. *J. Frankl. Inst.* **1998**, *6*, 1077–1086. [CrossRef]
19. Oldham, K.B. An introduction to the fractional calculus and some applications. In Proceedings of the Second International Workshop—Transform Methods and Special Functions, Varna, Bulgaria, 23–30 August 1996; pp. 598–609.
20. Liu, K.; Chen, Y.Q.; Zhang, X. An Evaluation of ARFIMA (Autoregressive Fractional Integral Moving Average) Programs. *Axioms* **2017**, *6*, 16. [CrossRef]
21. Pavlíčková, M.; Petráš, I. A note on time series data analysis using a fractional calculus technique. In Proceedings of the 15th International Carpathian Control Conference (ICCC), Velke Karlovice, Czech Republic, 28–30 May 2014; pp. 424–427.
22. Abramov, V. *Hurst Exponent Estimation*; Matlab Central File Exchange; MathWorks, Inc.: Natick, MA, USA, 2018. Available online: https://www.mathworks.com/matlabcentral/fileexchange/39069 (accessed on 23 April 2019).
23. Petráš, I. *Digital Fractional Order Differentiator/Integrator—FIR Type*; Matlab Central File Exchange; MathWorks, Inc.: Natick, MA, USA, 2003. Available online: http://www.mathworks.com/matlabcentral/fileexchange/3673 (accessed on 23 April 2019).
24. Oeters, F. *Metallurgy of Steelmaking*; Verlag Stahleisen mbH: Düsseldorf, Germany, 1994.
25. Turkdogan, E.T. *Fundamentals of Steelmaking*; Maney Publishing: Leeds, UK, 2010.
26. Kattenbelt, C.; Roffel, B. Dynamic Modeling of the Main Blow in Basic Oxygen Steelmaking Using Measured Step Responses. *Met. Mater. Trans. B* **2008**, *5*, 764–769. [CrossRef]

![mathematics logo] *mathematics*

MDPI

Article

Back to Basics: Meaning of the Parameters of Fractional Order PID Controllers

Inés Tejado [1,*], Blas M. Vinagre [1], José Emilio Traver [1], Javier Prieto-Arranz [1,2] and Cristina Nuevo-Gallardo [1]

[1]   Industrial Engineering School, University of Extremadura, 06006 Badajoz, Spain; bvinagre@unex.es (B.M.V.); jetraverb@unex.es (J.E.T.); Javier.PrietoArranz@uclm.es (J.P.-A.); cnuevog@unex.es (C.N.-G.)
[2]   School of Industrial Engineering, University of Castilla-La Mancha, 13071 Ciudad Real, Spain
[*]   Correspondence: itejbal@unex.es; Tel.: +34-924-289-300

Received: 8 May 2019; Accepted: 4 June 2019; Published: 11 June 2019

Abstract: The beauty of the proportional-integral-derivative (PID) algorithm for feedback control is its simplicity and efficiency. Those are the main reasons why PID controller is the most common form of feedback. PID combines the three natural ways of taking into account the error: the actual (proportional), the accumulated (integral), and the predicted (derivative) values; the three gains depend on the magnitude of the error, the time required to eliminate the accumulated error, and the prediction horizon of the error. This paper explores the new meaning of integral and derivative actions, and gains, derived by the consideration of non-integer integration and differentiation orders, i.e., for fractional order PID controllers. The integral term responds with selective memory to the error because of its non-integer order λ, and corresponds to the area of the projection of the error curve onto a plane (it is not the classical area under the error curve). Moreover, for a fractional proportional-integral (PI) controller scheme with automatic reset, both the velocity and the shape of reset can be modified with λ. For its part, the derivative action refers to the predicted future values of the error, but based on different prediction horizons (actually, linear and non-linear extrapolations) depending on the value of the differentiation order, μ. Likewise, in case of a proportional-derivative (PD) structure with a noise filter, the value of μ allows different filtering effects on the error signal to be attained. Similarities and differences between classical and fractional PIDs, as well as illustrative control examples, are given for a best understanding of new possibilities of control with the latter. Examples are given for illustration purposes.

Keywords: fractional; control; PID; parameter; meaning

1. Introduction

The proportional-integral-derivative (PID) controller is distinguished as the most common form of feedback. In process control today, more than 95% of the control loops are of PID type, but these controllers can be found in all areas where control is used.

Despite its straightforward structure, the popularity of PID controllers lies in the simplicity of the design procedures and in the effectiveness obtained to the system performance [1]. Those are the main reasons why PID controllers have survived many changes in technology, from mechanics and pneumatics to microprocessors via electronic tubes, transistors, integrated circuits, among others. Actually, practically all PID controllers made today are based on microprocessors, so this electronic element has had a dramatic influence on this kind of control providing PIDs additional advances features, such as gain scheduling, continuous adaptation, and automatic tuning [2].

From the control engineering point of view, improving system behavior is the major concern. To that end, the generalization of classical PID controllers to non-integer orders of integration and differentiation was firstly proposed in [3]. Intuitively, with this extension of classical PIDs there are more tuning parameters and, consequently, more flexibilities in adjusting time and frequency responses of the control system. This also translates in more robustness in designs.

However, the first step when applying an existing or new controller is to understand exactly what their actions can do in closed-loop in order to take full advantage of the possible effects on the system response. In the case of integer order, the interpretation of the three actions of PIDs seems to be clear [4–6]:

- the proportional action is simply proportional to the current control error;
- the integral action is related to the past values of the control error, so represents the accumulated error, i.e., the area under the error curve;
- the derivative action predicts future values of the error or, in other words, corrects based on the rate of change of the deviation from the set-point.

Since the pioneering work of Podlubny, there is ample evidence that supports that fractional order PIDs (FOPIDs) have been extensively studied and applied in many fields. Undoubtedly, the lists that are reported below are quite far from aiming at completeness, failing to mention literal hundreds of other published texts related to FOPIDs; relevant and recent papers were selected for giving the reader an idea of the development volume on this topic that can be found in the specialized literature. Fundamentals of FOPIDs can be found in, e.g., [7–10]. In what concerns design methods, among the reported, the following can be highlighted: Ziegler–Nichols-type rules [11–14], optimal tuning [15–18], tuning for robustness purposes [19–21], auto-tuning [22,23], and tuning based on reducing the number of parameters [24]. Likewise, numerous advanced control schemes based on FOPID controllers were proposed, such as, Smith predictors structures [25–27], internal mode control [28–30], hybrid control [31], gain scheduling [32–34], gain and order scheduling [35,36], among others. Current reviews in the development of FOPIDs can be found in [37–42]; likewise, few current examples of application in industry are described in [43–50].

Despite so many variations and applications of the FOPID algorithm, as well as design and tuning methods, up to now a detailed description of the meaning of the parameters of such controllers cannot be found in the literature. Nevertheless, although few, studies on the geometric and physical interpretation of integrals and derivatives of arbitrary (not necessary integer) order have been published [51,52], but it still remains as an open problem.

These circumstances make the understanding of the meaning of the parameters of FOPIDs a priority. With so much in play, this paper explores the new meaning of integral and derivative actions, and gains, derived by the consideration of non-integer integration and differentiation orders. Similarities and differences between classical and fractional PIDs, as well as illustrative control examples, are given for a best understanding of new possibilities of control with the latter.

The remainder of this paper is organized as follows. Section 2 describes the control algorithm of classical PID and its generalization to non-integer order. Section 3 explains the meaning of the parameters of non-integer PIDs. Section 4 discusses similarities and differences between classical and fractional PIDs. Illustrative examples are given in Section 5. Main conclusions of this paper are drawn in Section 6.

2. Generalities

This section describes the control algorithm of classical PIDs and its generalization to non-integer order.

2.1. Classical PID Controller

The use of PID control consists of applying properly the combination of three types of corrective actions to the error signal, which represents how far or near is the desired output from the actual output. As widely known, these three control actions are proportional, integral and derivative.

The key aspect when tuning PID controllers is in deciding how to best combine those three terms to achieve the most efficient regulation of the process variable for the considered problem. As well known, the most obvious way is to use a simple weighted sum where each term is multiplied by a tuning constant or gain, and the results are then added together as follows:

$$u(t) = K_p \left(e(t) + \frac{1}{T_i} \int_0^t e(\tau)d\tau + T_d \frac{de(t)}{dt} \right), \tag{1}$$

where $u(t)$ is the control signal, $e(t)$ is the control error ($e = y_{sp} - y$, i.e., the difference between the desired set-point, y_{sp}, and the measured process variable, y), and K_p, T_i, and T_d are the controller parameters: proportional gain, integral time constant, and derivative time constant, respectively.

Control law (1) guarantees that the present (due to the proportional action), the past (by means of the integral action) and the future of the error (by the derivative action) are taken into account, as shown in Figure 1. Two main observations can be made to (1): (1) the controller needs only compute the current error between the measured process variable and the desired set-point to calculate how much and how fast that difference has been changing over time, and (2) the relative contributions of each term then can be then adjusted by choosing appropriate values of the controller parameters.

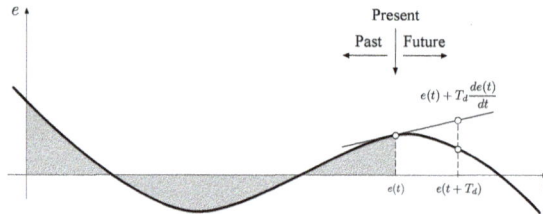

Figure 1. Evolution of the error (classical proportional-integral-derivative (PID) controllers).

2.2. Fractional Order PID Controller

The generalization to non-integer orders of (1) leads to the typical algorithm of a FOPID controller, i.e.,

$$u(t) = K_p \left(e(t) + \frac{1}{T_i} \mathcal{D}^{-\lambda} e(t) + T_d \mathcal{D}^{\mu} e(t) \right), \tag{2}$$

where and $\lambda, \mu \in \mathbb{R}^+$ are the non-integer orders of the integral and derivative terms, respectively, and $\mathcal{D}$ is the fractional operator defined as Riemann–Liouville as [10,53]

$$\mathcal{D}^{-n} f(t) = \frac{1}{\Gamma(n)} \int_0^t f(y)(t-y)^{n-1} dy, \tag{3}$$

(n is a general non-integer order, and $\Gamma(-)$, the gamma function).

Similarly to the classical version, control law (2) combines the three natural ways of taking into account the error: current, accumulated, and predicted error. However, in contrast to the integer counterpart, fractional operators are non local, which results in new meaning of the integral and derivative actions, and gains. In particular, the fractional integration of the error is not already the area under its curve; it can be viewed as the area of the projection of the curve onto a plane. Hence, what the integration order λ is doing is a selection of the history of the error or, what is the same, the integral term responds with selective memory to the past values of the error. For its part, the action of a controller with proportional and fractional derivative action may be interpreted as if the control is made proportional to the predicted process output, where the prediction is definitively different from the classical case: it is made by extrapolating the error by a straight line that is not tangent to the error curve at the current value of the error, or by a curve (i.e., linear and non-linear extrapolations). More details of these and other possibilities will be explained in Section 3.

Figure 2 is an attempt to illustrate these effects. It should be remarked that, in this figure, the error was considered as the function $e(t) = t + 0.5 \sin(t)$ (taken from [51]). To show the integral and derivative effects for non-integer orders, it must be said that:

- For integral action, left-side Riemann-Liouville fractional integral of the error was considered as:

$$_0I_t^\lambda e(t) = \int_0^t e(\tau) dg_t(\tau),$$ (4)

where

$$g_t(\tau) = \frac{1}{\Gamma(\lambda+1)} \left(t^\lambda - (t-\tau)^\lambda \right).$$ (5)

Then, taking the axes τ, g_t and e, function $g_t(\tau)$ was plotted in the plane (τ, g_t) for $0 \le \tau \le t$. Along the obtained curve, a "fence" was plotted varying $e(\tau)$, so the top edge of the "fence" is a 3D line $(\tau, g_t(\tau), e(\tau))$ for $0 \le \tau \le t$. The area of the projection of this "fence" onto the plane (τ, e) corresponds to the value of the integer order integral of the error (i.e., the classical area under the curve), whereas that of the projection onto the plane (g_t, e) corresponds to the value of fractional order integral (4). All this reasoning was adapted from [51].

- For derivative action, straight lines that pass through the point $e(t)$ and whose slope matches the value of the differentiation of order μ of the error curve at that point were plotted.

The following values of the orders were taken for simulations: $\lambda = 0.75$, $\mu_1 = 1.4$, $\mu_2 = 1$, and $\mu_3 = 0.4$.

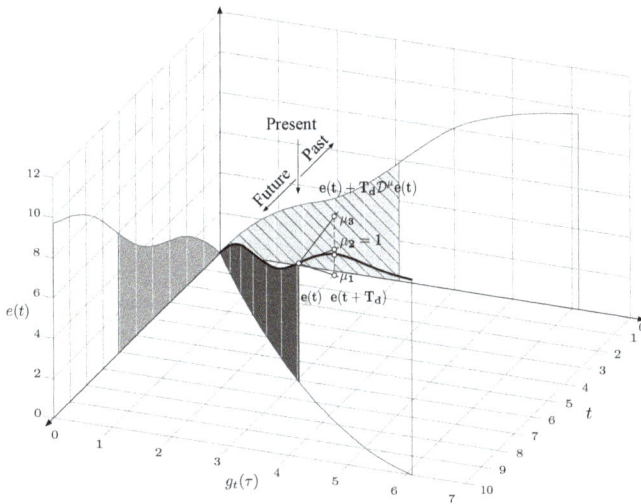

Figure 2. Evolution of the error (fractional PID controllers).

3. Going Into Detail About Parameters

This section contains the explanation of the meaning of the parameters of FOPID controllers, in comparison with those classical of integer order. It should be remarked that, despite the fact the proportional action is not affected by a fractional order, it is also included for completeness of information.

3.1. Proportional Action

The proportional action increases the deviation between the set-point value y_{sp} and the system output y (i.e., the error) with the proportional gain K_p. The main drawback of using a pure proportional control action is that it produces a steady-state error, which motivates that it can be also considered as:

$$u(t) = K_p e(t) + u_b, \tag{6}$$

where u_b is a bias or reset [4]. Indeed, when $e = 0$, the control signal reduces to $u(t) = u_b$. The parameter u_b usually takes the value $(u_{max} + u_{min})/2$, where u_{max} and u_{min} are the maximum and minimum limits of the actuator, respectively. However, sometimes u_b can be adjusted manually to a value that ensures that the steady-state error is zero at a given set-point.

Likewise, the proportional gain can be specified in terms of its inverse proportional band, P_b, which represents the percentage of change in the error signal necessary to cause a full-scale change in the proportional action. As can be observed in Figure 3, the tendency of y to oscillate increases as P_b decreases. The large oscillations occurring with a small P_b are due to the fact that the power is reduced very quickly when the system output enters the proportional band, meaning a balanced state cannot be established immediately.

Indeed, confusing "proportional band" with "proportional gain" leads to a decreased proportional action when the control engineer wants more, and vice-versa.

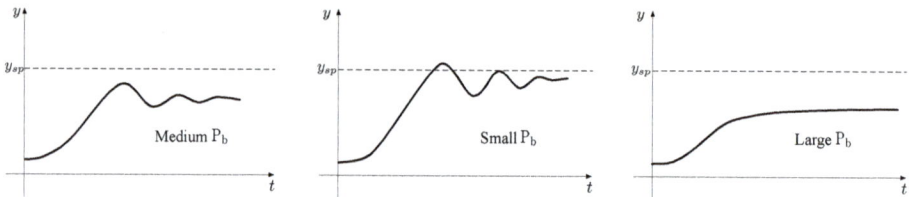

Figure 3. System response for different P_b.

3.2. Integral Action

The main function of the integral action is to guarantee that the steady-state error is zero, i.e., the output of the controlled system is equal to the desired set-point in steady-state. The following simple explanation proves this affirmation. Let us consider a system controlled by a fractional order proportional-integral (PI) controller in steady-state where both the control signal (u_{ss}) and the error (e_{ss}) are constant. If this is the case, the control signal will be given by [3]

$$u_{ss} = K_p \left(e_{ss} + \frac{e_{ss}}{T_i} \frac{t^\lambda}{\Gamma(\lambda+1)} \right). \tag{7}$$

While $e_{ss} \neq 0$, this clearly contradicts the hypothesis that the control signal u_{ss} is constant. Thus, similarly to the integer case, a controller with integral action of fractional order will always give zero error in steady-state.

Integral action was known originally as a device that automatically resets the bias u_b of a proportional controller. Figure 4 shows the scheme of the extension to non-integer orders of a PI controller with automatic reset. As can be observed, the first order filter in the feedback loop is replaced by its fractional version of order λ. From the block diagram, the control signal can be obtained as

$$u = K_p e + I, \tag{8}$$

with

$$I = U(s)\frac{1}{1 + T_i s^\lambda}$$

$$\Rightarrow I + T_i \frac{d^\lambda I}{dt} = e,$$

$$\Rightarrow T_i \mathcal{D}^\lambda I = K_p e$$

$$\Rightarrow I = K_p \frac{1}{T_i} \mathcal{D}^{-\lambda} e(t). \tag{9}$$

Substituting (9) in (8), the control signal is given by

$$u(t) = K_p \left(e(t) + \frac{1}{T_i} \mathcal{D}^{-\lambda} e(t) \right), \tag{10}$$

thus the control scheme of the figure is equivalent to a fractional order PI controller. Figure 5 shows unit step responses of the fractional system in the feedback loop in Figure 4. Unlike the integer order case (see Figure 5a), both the velocity and the shape of resetting can be controlled by the integration order, λ, as shown in Figure 5b. It can be observed that it is even possible to obtain underdamped responses when $\lambda > 1$.

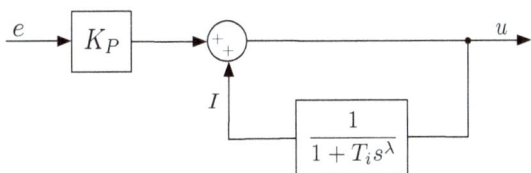

Figure 4. Fractional proportional-integral (PI) controller scheme (classical implementation with automatic reset).

It is important to remark that function `fode_sol()` was taken from [10] to obtain the step responses in MATLAB.

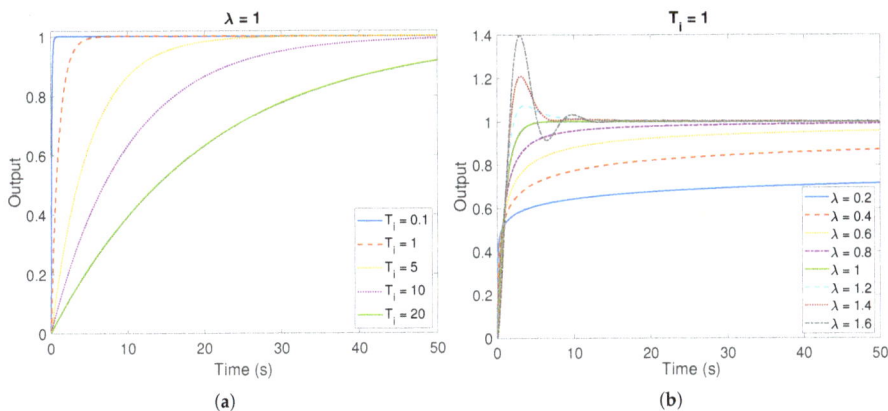

Figure 5. Effects of the automatic reset of the integral action when changing: (a) integral time constant T_i (integer case) (b) integration order λ (fractional case).

3.3. Derivative Action

The objective of the derivative action is to improve the system stability in closed-loop. Classical derivative controllers give responses to changing error signals but do not, however, respond to constant error signals, since with a constant error the rate of change of error with time is zero. Because of this, the derivative term is combined with, at least, the proportional term. By contrast, fractional derivative controllers do response to constant error signals; the differentiation of fractional order of a constant is different from zero.

The derivative action of a proportional-derivative (PD) controller can be viewed as a crude prediction of the error in future, where the prediction is made by extrapolating the error by the tangent to the error curve at time t, being T_d the prediction horizon (see Figure 1). (Actually, the derivative action uses linear extrapolation, not prediction.) For the fractional case, the basic structure of the controller is

$$u(t) = K_p\left(e(t) + T_d \mathcal{D}^\mu e(t)\right). \tag{11}$$

Analogously to the classical case, an approximation of $e(t + T_d)$ may be

$$e(t + T_d) \approx e(t) + T_d \mathcal{D}^\mu e(t). \tag{12}$$

The control signal is then proportional to an estimation of the error at time T_d ahead over a straight line that, in general, is not tangent to the error curve, and whose slope matches the value of the differentiation of μ-th order of the error curve at the point $e(t)$ [52], as shown in Figure 2. Likewise, another way of viewing the prediction is from fractional Taylor series for fractional derivatives. In this case, the prediction horizon is done over a curve (i.e., non-linear extrapolation). More details about prediction can be found in Appendix A.

Therefore, different prediction horizons (in fact, linear and non-linear extrapolations) for the error can be obtained choosing accordingly the value of μ.

The classical implementation of a fractional derivative action is illustrated in Figure 6. From this figure, the following relation can be obtained

$$U(s) = \left(1 - \frac{1}{1 + T_d s^\mu}\right) E(s) = \frac{T_d s^\mu}{1 + T_d s^\mu} E(s), \tag{13}$$

which corresponds to a derivative action of fractional order with noise filter. It is well known that the derivative part of the PID controller requires low-pass filtering to limit the high-frequency gain.

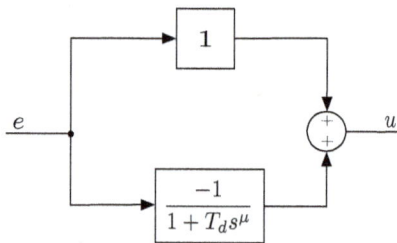

Figure 6. Fractional order derivative action scheme (classical implementation).

Let us focus only on the noise filter of (13), i.e.,

$$G_f(s) = \frac{1}{1 + T_d s^\mu}. \tag{14}$$

Notice that: (1) it has the property $G_f(0) = 1$ to ensure that the process output equals the set-point in steady-state; (2) as usual, it has a low-pass character; and (3) the order of the filter is μ.

In the literature, the classical filters to reduce noise effects in PID controllers are up to second order, usually given as follows [54]:

$$G_{f1}(s) = \frac{1}{1 + T_f s}, \tag{15}$$

$$G_{f2}(s) = \frac{1}{1 + T_f s + T_f^2 / 2s^2}, \tag{16}$$

where T_f is the filter time constant. Note that filter (16) has complex poles with damping ratio $\delta = 0.707$.

Figure 7 shows Bode plots of the filters, of both integer and fractional order, for $T_d = T_f = 0.1$. As can be seen in Figure 7b, changes in the order μ allow to have different frequency responses ranged between those of the two integer filters (Figure 7a), and consequently, different filtering effects on the error signal can be attained. Another issue to take into account is that most design methods for classical PIDs do not consider noise. Due to this fact, the filter time constant is often suggested to be chosen as a fraction of the derivative time, i.e., $T_f = T_d / N$. However, this solution has severe drawbacks, as reported in [55]. On the contrary, this is not necessary for the fractional case because there is an only parameter to be tuned, μ, and it is included in the controller design.

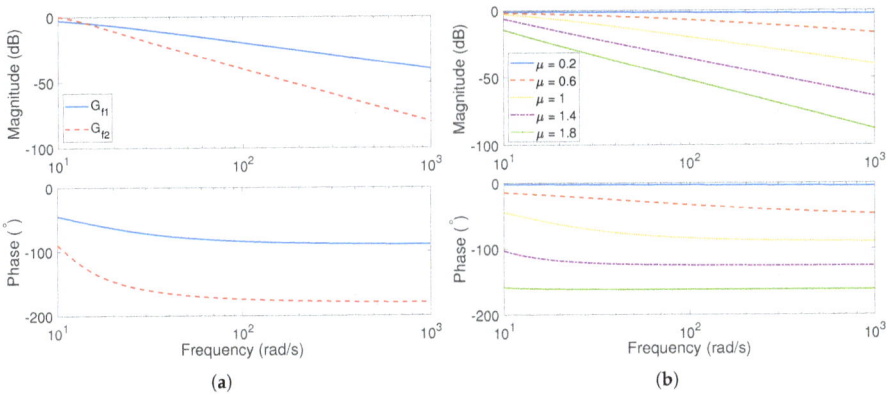

Figure 7. Frequency response of the noise filter: (**a**) first and second order low-pass filters (integer case) (**b**) fractional filter for different values of μ (fractional case). For simulations, the following values were taken: $T_d = T_f = 0.1$.

4. Classical Versus Fractional PIDs

This section discusses the similarities and differences between classical and fractional PIDs. In particular, only integral and derivatives parts are compared— the proportional part is similar for both kinds of controllers. The properties described below are summarized in Table 1.

Table 1. Similarities and differences between classical and fractional proportional-integral-derivatives (PID) controllers.

Action	Domain	Effect on	Integer	Fractional
I		Steady-state error	Elimination	
	Time	u with respect to the sign of e	If $e > 0$, u grows linearly with time	The same but the growing or the decrease is not linear with time
			If $e < 0$, u decreases linearly with time	
		Automatic reset	The velocity of reset can be changed	Both the velocity and the shape of reset can be changed
	Frequency	Frequency response	The magnitude curve decreases with a slope of 20 dB/dec	The magnitude curve decreases with a slope of 20λ dB/dec
			Decrease of $\pi/2$ rad in the phase curve	Decrease of $(\pi\lambda)/2$ rad in the phase curve
D	Time	Constant errors	Does not response, so the derivative term needs to be combined with, at least, the proportional term	Does response, so the derivative term can be used individually
		Prediction horizon	Time T_d ahead over the tangent to the error curve at $e(t)$	Time T_d ahead over a curve (non-linear extrapolation) or over a straight line (linear extrapolation) that passes through the point $e(t)$ and whose slope matches the value of the fractional differentiation of order μ of the error curve at that point
	Frequency	Frequency response	The magnitude curve grows with a slope of 20 dB/dec	The magnitude curve grows with a slope of 20λ dB/dec
			Increment of $\pi/2$ rad in the phase curve	Increment of $(\pi\lambda)/2$ rad in the phase curve
	Frequency		Low-pass filters up to second order	Low-pass filter of order μ
		Filtering	Usually needs two parameters to be tuned, i.e., T_f (filter time constant) and N (ratio between T_d and T_f)	Parameter μ allows to have different frequency responses, ranged between those of the two integer filters, and consequently, have different filtering effects on the error signal.

4.1. Integral Part

As is well known, the main effects of the integral action are those that make the system response slower, decrease its relative stability, and eliminate the steady-state error for inputs for which previously had a finite error. These effects can be observed in the different domains of analysis as follows. In the time-domain, the effects on the transient response cause a decrease of the rising time and an increase of the settling time and the overshoot. In the complex plane, the integral action causes a displacement of the root locus of the system towards the right half-plane. Finally, in the frequency-domain, an increase of -20 dB/dec in the slopes of the magnitude curves and a decrement of $\pi/2$ in the phase plots can be observed.

In the case of a fractional integration order $\lambda \in (0,1)$, the selection of the value of λ translates into a certain weighting of the effects mentioned above. In the time-domain, for example, due to the fact that the integral action only responds to errors other than zero by increasing the control action, for positive errors, and decreasing it in case of negative, if the error is constant, the control action can be ramped up with different slopes or velocities, as can be observed in Figure 8a. For a square error signal (Figure 8b), it can be observed that there is a set of effects of the control action over the error that range from the pure proportional action ($\lambda = 0$) to the classical integral action ($\lambda = 1$). For intermediate values of λ, the control action grows when the error is constant, which results in a removal of the steady-state error, whereas decreases when the error goes to zero, which reduces the instability of the system. In the complex plane, it can be seen that the selection of a value of parameter λ moves the root locus of the system towards the right half-plane. In the frequency-domain, the fact that λ can vary between 0 and 1 introduces the possibility of adding a constant increment to the slope of the magnitude curve between 0 and -20 dB/dec (actually, a value of -20λ dB/dec), as well as a constant lag to the phase curve between 0 and $-\pi/2$ rad (specifically, a value of $-(\pi\lambda)/2$ rad).

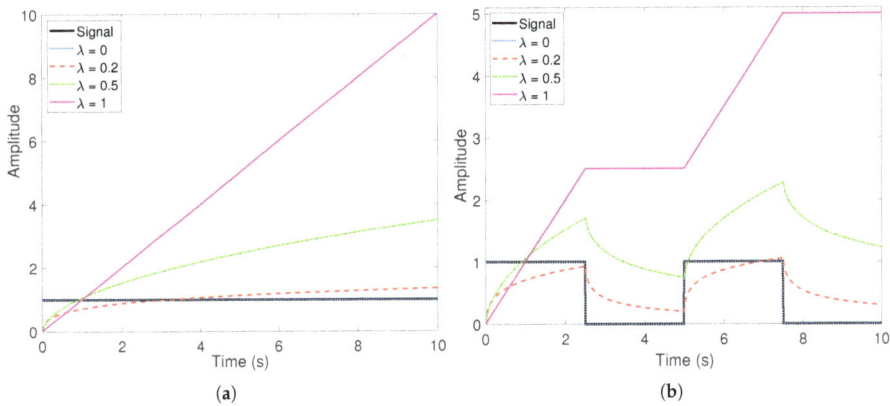

(a)

(b)

Figure 8. Effects of the integral action according to integration order λ on different error signals: (**a**) constant (**b**) square.

4.2. Derivative Part

The derivative action increases the system stability and tends to accentuate the noise effects at high frequencies. In the time-domain, this is manifested, mainly, by a decrease in both the overshoot and the settling time. In the complex plane, it produces a displacement of the system root locus towards the left half-plane. In the frequency-domain, a constant phase lead of $\pi/2$ and an increase of 20 dB/dec in the slopes of the magnitude curves.

Following a reasoning parallel to that made for the integral action, it is easy to understand that all these effects are weighted by choosing an appropriate value of the order of the derivative, $\mu \in (0,1)$ (see Figure 9).

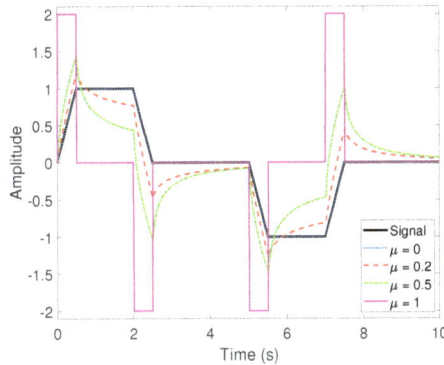

Figure 9. Effects of the derivative action according to differentiation order μ on a trapezoidal signal.

5. Illustrative Examples

This section provides two examples to illustrate the properties of both integral and derivative actions, as well the effects of their fractional orders.

Example 1. *Let us consider a system given by the following transfer function [4]*

$$G(s) = \frac{1}{(s+1)^3} \tag{17}$$

19

controlled by a fractional order PI of the form:

$$C(s) = K_p \left(1 + \frac{1}{T_i s^\lambda} \right) \tag{18}$$

To illustrate the properties of the integral action and the effect of its fractional order, Figure 10 shows unit step responses in closed-loop of system (17) controlled by fractional PI controller (18). The proportional gain is constant, $K_p = 1$, whereas the integral time T_i and λ are changed individually in Figure 10a,b, respectively. More precisely, responses for the classical case ($\lambda = 1$) are plotted in Figure 10a. As expected, the steady-state error is removed when T_i takes finite values; the case $T_i = \infty$ corresponds to pure proportional control, where the steady-state error is 50 percent. Likewise, the smaller the values of T_i, the faster the response, but the more oscillatory [4]. The effects of changing the integration order, λ, can be seen in Figure 10b (T_i is fixed to 2). In this case, the smaller the values of λ, the slower the response. Unlike T_i, oscillations and system response are not affected by parameter λ.

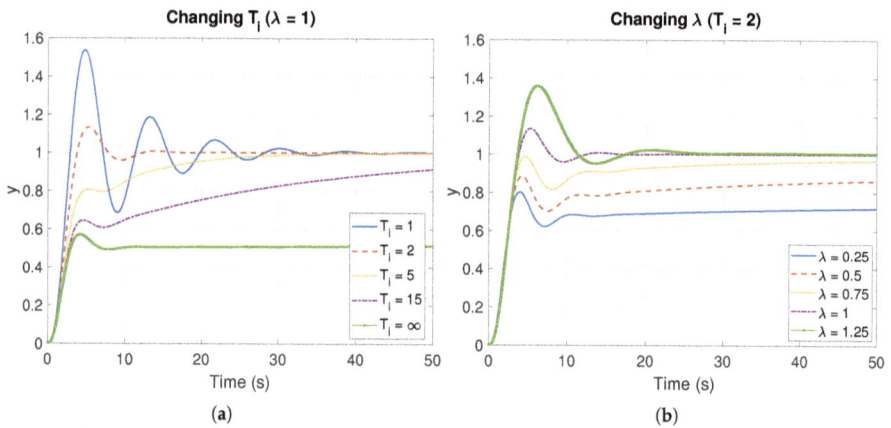

Figure 10. Step responses in closed-loop of system (17) controlled by fractional PI controller (18) when changing: (**a**) integral time constant T_i ($K_p = 1$, $\lambda = 1$) (**b**) integration order λ ($K_p = 1$, $T_i = 2$).

Example 2. *Now, consider the double integrator with unit gain, i.e.,*

$$G(s) = \frac{1}{s^2} \tag{19}$$

controlled by a fractional order PD of the form:

$$C(s) = K_p \left(1 + T_d s^\mu \right). \tag{20}$$

The properties of the derivative action and its non-integer order are illustrated in Figure 11, which shows unit step responses in closed-loop of system (19) controlled by fractional PD controller (20). Similarly to the previous example, the proportional gain is constant, $K_p = 1$, whereas the derivative time T_d and μ are changed in an individual way in Figure 11a,b, respectively. In particular, responses for the classical case ($\mu = 1$) are plotted in Figure 11a. As expected, damping increases when T_d increases. In other words, the higher the values of T_d, the faster the response. The effects of changing μ can be observed in Figure 11b (T_d is set to 2). In this case, the smaller the values of μ, the slower the damping. Unlike T_d, μ only affects the overshoot.

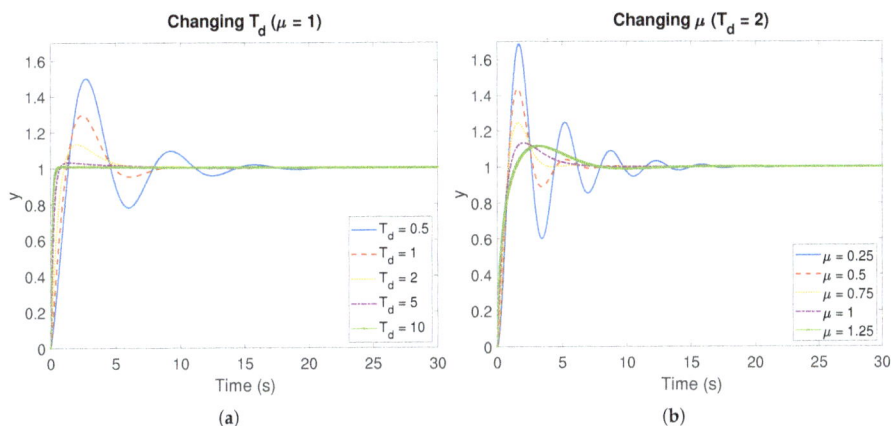

Figure 11. Step responses in closed-loop of system (19) controlled by fractional PD controller (20) when changing: (**a**) derivative time constant T_i ($K_p = 1$, $\mu = 1$) (**b**) integration order μ ($K_p = 1$, $T_d = 2$).

It must be said that the function `fode_sol()` was also used to obtain the step responses of both examples.

6. Conclusions

Although ample studied and successfully applied to many kinds of control problems, the extension of classical proportional-integral-derivative (PID) controllers to non-integer orders presents a main weakness: basis, in terms of understanding of the effects of their parameters on the system response, is sometimes omitted, and even unknown. This can be explained, in part, because the geometric and physical interpretation of integrals and derivatives of arbitrary (not necessary integer) order still remains as an open problem. This paper has focused on the new meaning of integral and derivative actions, and gains, derived by the consideration of non-integer integration and differentiation orders, i.e., for fractional order PID (FOPID) controllers. Similarities and differences between classical and fractional PIDs, as well as illustrative examples were given for a best understanding of the possibilities of control with the latter.

When the integral term is concerned, it was shown that it responds with selective memory because of the non-integer order λ. That was also viewed as the area of the projection of the error curve onto a plane, which is definitively different from the classical area under the error curve. Moreover, taking into account a fractional proportional-integral (PI) controller scheme with automatic reset, unlike the integer order case, it was also shown that both the velocity and the shape of resetting can be controlled by the integration order, λ.

With respect to derivative action, it was illustrated that different prediction horizons (in fact, linear and non-linear extrapolations) for the error can be obtained for a fractional proportional-derivative (PD) choosing accordingly the value of the differentiation order, μ. Likewise, in case of an structure with noise filter, the value of μ allows different filtering effects on the error signal to be attained.

Author Contributions: This work involved all coauthors. I.T. wrote the original draft and contributed to the investigation and the analysis, edited the manuscript and contributed to the illustrations and examples. B.M.V. conceived the idea, contributed to the editing and supervised all the manuscript. J.E.T. wrote the plotting software and contributed to the illustrations. J.P.-A. and C.N.-G. contributed to the illustrations and the editing.

Funding: This research has been supported in part by the Spanish Agencia Estatal de Investigación (AEI) under Project DPI2016-80547-R (Ministerio de Economía y Competitividad), and in part by the European Social Fund (FEDER, EU).

Conflicts of Interest: The authors declare no conflict of interest.

Appendix A. Analysis of Prediction Horizon

This appendix gives information about the approximation of the prediction of the error at time T_d ahead, i.e., the approximation of $e(t + T_d)$ to understand the meaning of a fractional order derivative action.

Given a continuous function y, the fractional Euler method establishes that [56]

$$y(t_{i+1}) \approx y(t_i) + \mathcal{D}^\mu y(t_i) \frac{h^\mu}{\Gamma(\mu+1)} \left((i+1)^\mu - i^\mu\right). \tag{A1}$$

Let assume that the error $e(t)$ is a continuous function, and $T_d = kh$, $k \in \mathbb{Z}^+$, where h is a fixed step. Thus, $e(t + T_d)$ can be approximated by:

$$e(t + T_d) \approx e(t) + \mathcal{D}^\mu e(t) \frac{h^\mu}{\Gamma(\mu+1)} \left(T_d^\mu - t^\mu\right), \tag{A2}$$

which is referred to as approximation #1. Likewise, analogously with the integer case, the prediction may be also expressed as

$$e(t + T_d) \approx e(t) + T_d \mathcal{D}^\mu e(t), \tag{A3}$$

It is referred to as approximation #2.

Hence, taking into account approximation #1, the prediction in future of the error is done from a non-linear extrapolation due to the fact that the term on the right in (A2) is non-linear. In contrast, the prediction with (A3) is carried out over a straight line (linear extrapolation) that passes for point t, and whose slope matches the value of the μ-th order differentiation of the error. Thus, the control signal is proportional to an estimation of the control error at time T_d ahead, where that estimation is obtained by non-linear or linear extrapolation. Both predictions are illustrated in Figure A1 for different values of μ and T_d, namely, $\mu = \{0.4, 0.8, 1, 1.25\}$ and $T_d = \{1, 2, 1.5\}$, and considering the error as $e(t) = t + 0.5 \sin(t)$ and $e(t) = 0.01t^2(1 - 2\cos(t))$. As can be observed, although there are big differences between the approximations above, the prediction function assumed to the derivative term is guaranteed at time T_d ahead, especially when approximation given by (A2) is taken: it approaches, to a large extent, the actual curve of the error.

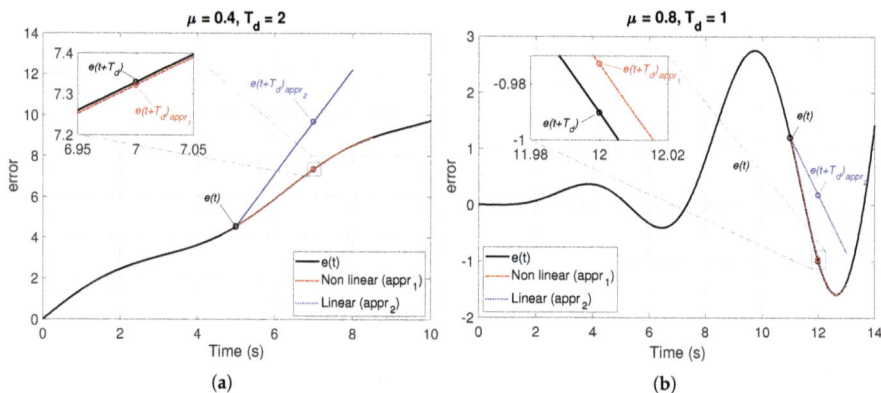

Figure A1. *Cont.*

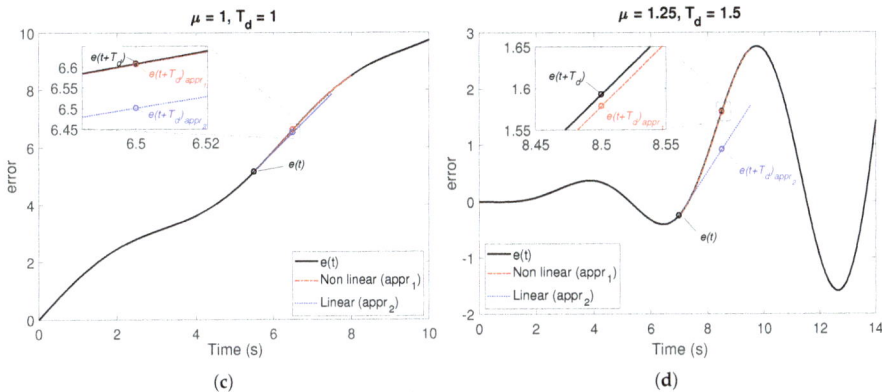

Figure A1. Prediction horizons of derivative part depending on the order μ: (**a**) $\mu = 0.4$, $T_d = 2$ (**b**) $\mu = 0.8$, $T_d = 1$ (**c**) $\mu = 1$, $T_d = 1$ (**d**) $\mu = 1.25$, $T_d = 1.5$. Notice that cases (**a**) and (**c**) correspond to the error as the function $e(t) = t + 0.5\sin(t)$, whereas the error is given by $e(t) = 0.01t^2(1 - 2\cos(t))$ in cases (**b**) and (**d**).

References

1. Aström, K.J.; Hägglund, T. *PID Controllers: Theory, Design and Tuning*, 2nd ed.; ISA—The Instrumentation, Systems, and Automation Society: Research Triangle Park, NC, USA, 1995.
2. Aström, K.J.; Murray, R.M. *Feedback Systems: An Introduction for Scientists and Engineers*, 2nd ed.; Princeton University Press: Princeton, NJ, USA, 2008.
3. Podlubny, I. Fractional-order systems and PI$^\Lambda$D$^\mu$-controllers. *IEEE Trans. Autom. Control* **1999**, *44*, 208–214. [CrossRef]
4. Aström, K.J.; Hägglund, T. *Advanced PID Control*; Chapter PID Control; ISA—The Instrumentation, Systems, and Automation Society: Research Triangle Park, NC, USA, 2006; pp. 64–94.
5. Visioli, A. *Practical PID Control*; Advances in Industrial Control; Chapter Basics of PID Control; Springer: Berlin/Heidelberg, Germany, 2006; pp. 1–18.
6. Li, Y.; Heong, K.; Chong, G.C.Y. PID Control System Analysis and Design. Problems, Remedies, and Future Directions. *IEEE Control Syst. Mag.* **2006**, *26*, 32–41.
7. Vinagre, B.M.; Feliu-Batlle, V.; Tejado, I. Fractional Control: Fundamentals and User Guide. *Revista Iberoamericana de Automática e Informática Industrial* **2016**, *13*, 265–280. [CrossRef]
8. Valério, D.; Sá da Costa, J. *An Introduction to Fractional Control*; IET Control Engineering, The Institution of Engineering and Technology: Stevenage, UK, 2013.
9. Vinagre, B.M.; Monje, C.A. *PID Control in the Third Millennium. Lessons Learned and New Approaches*; Advances in Industrial Control; Chapter Fractional-Order PID; Springer: Berlin/Heidelberg, Germany, 2012; pp. 465–494.
10. Monje, C.A.; Chen, Y.Q.; Vinagre, B.M.; Xue, D.; Feliu, V. *Fractional-Order Systems and Controls. Fundamentals and Applications*; Springer: Berlin/Heidelberg, Germany, 2010.
11. Lorenzini, C.; Bazanella, A.S.; Alves Pereira, L.F.; Goncalves da Silva, G.R. The generalized forced oscillation method for tuning PID controllers. *ISA Trans.* **2019**, *87*, 68–87. [CrossRef] [PubMed]
12. Tajjudin, M.; Rahiman, M.H.F.; Arshad, N.M.; Adnan, R. Robust Fractional-Order PI Controller with Ziegler-Nichols Rules. *Int. J. Electr. Comput. Eng.* **2013**, *7*, 1034–1041.
13. Gude, J.J.; Kahoraho, E. Modified Ziegler-Nichols method for fractional PI controllers. In Proceedings of the 2010 IEEE 15th Conference on Emerging Technologies & Factory Automation (ETFA 2010), Bilbao, Spain, 13–16 September 2010.
14. Valério, D.; Sá da Costa, J. Tuning of fractional PID controllers with Ziegler-Nichols-type rules. *Signal Process.* **2006**, *10*, 2771–2784. [CrossRef]

15. Hekimoglu, B. Optimal Tuning of Fractional Order PID Controller for DC Motor Speed Control via Chaotic Atom Search Optimization Algorithm. *IEEE Access* **2019**, *7*, 38100–38114. [CrossRef]
16. Kesarkar, A.A.; Selvaganesan, N. Tuning of optimal fractional-order PID controller using an artificial bee colony algorithm. *Syst. Sci. Control Eng.* **2015**, *3*, 99–105. [CrossRef]
17. Padula, F.; Visioli, A. Tuning rules for optimal PID and fractional-order PID controllers. *J. Process Control* **2011**, *21*, 69–81. [CrossRef]
18. Biswas, A.; Das, S.; Abraham, A.; Dasgupta, S. Design of fractional-order $PI^{\lambda}D^{\mu}$ controllers with an improved differential evolution. *Eng. Appl. Artif. Intell.* **2009**, *22*, 343–350. [CrossRef]
19. Liu, L.; Zhang, S. Robust Fractional-Order PID Controller Tuning Based on Bode's Optimal Loop Shaping. *Complexity* **2018**, *2018*, 6570560. [CrossRef]
20. HosseinNia, S.H.; Tejado, I.; Vinagre, B.M. A Method for the Design of Robust Controllers Ensuring the Quadratic Stability for Switching Systems. *J. Vib. Control* **2014**, *20*, 1085–1098. [CrossRef]
21. Monje, C.A.; Calderón, A.J.; Vinagre, B.M.; Chen, Y.Q.; Feliu, V. On fractional PI^{λ} controllers: Some tuning rules for robustness to plant uncertainties. *Nonlinear Dyn.* **2004**, *38*, 369–381. [CrossRef]
22. De Keyser, R.; Muresan, C.I.; Ionescu, C.M. A novel auto-tuning method for fractional order PI/PD controllers. *ISA Trans.* **2016**, *62*, 268–275. [CrossRef] [PubMed]
23. Monje, C.A.; Vinagre, B.M.; Feliu, V.; Chen, Y.Q. Tuning and auto-tuning of fractional order controllers for industry applications. *Control Eng. Pract.* **2008**, *16*, 798–812. [CrossRef]
24. Chevalier, A.; Francis, C.; Copot, C.; Ionescu, C.M.; De Keyser, R. Fractional-order PID design: Towards transition from state-of-art to state-of-use. *ISA Trans.* **2019**, *84*, 178–186. [CrossRef] [PubMed]
25. Feliu-Batlle, V.; Rivas-Perez, R.; Castillo-Garcia, F.J. Simple Fractional Order Controller Combined with a Smith Predictor for Temperature Control in a Steel Slab Reheating Furnace. *Int. J. Control Autom. Syst.* **2013**, *11*, 533–544. [CrossRef]
26. Feliu-Batlle, V.; Rivas Pérez, R.; Castillo García, F.J.; Sánchez Rodríguez, L. Smith predictor based robust fractional order control: Application to water distribution in a main irrigation canal pool. *J. Process Control* **2009**, *19*, 506–519. [CrossRef]
27. Luan Vu, T.N.; Lee, M. Smith predictor based fractional-order PI control for time-delay processes. *Korean J. Chem. Eng.* **2014**, *31*, 1321–1329.
28. Lakshmanaprabu, S.K.; Banu, U.S.; Hemavathy, P.R. Fractional order IMC based PID controller design using Novel Bat optimization algorithm for TITO Process. *Energy Proc.* **2017**, *117*, 1125–1133. [CrossRef]
29. Muresan, C.I.; Dutta, A.; Dulf, E.H.; Pinar, Z.; Maxim, A.; Ionescu, C.M. Tuning algorithms for fractional order internal model controllers for time delay processes. *Int. J. Control* **2016**, *89*, 579–593. [CrossRef]
30. Maamar, B.; Rachid, M. IMC-PID-fractional-order-filter controllers design for integer order systems. *ISA Trans.* **2014**, *53*, 1620–1628. [CrossRef] [PubMed]
31. HosseinNia, S.H.; Tejado, I.; Vinagre, B.M.; Milanés, V.; Villagrá, J. Experimental Application of Hybrid Fractional Order Adaptive Cruise Control at Low Speed. *IEEE Trans. Control Syst. Technol.* **2014**, *22*, 2329–2336. [CrossRef]
32. Ahmed, M.F.; Dorrah, H.T. Design of gain schedule fractional PID control for nonlinear thrust vector control missile with uncertainty. *Automatika* **2018**, *59*, 357–372. [CrossRef]
33. Tejado, I.; Milanés, V.; Villagrá, J.; Vinagre, B.M. Fractional Network-based Control for Vehicle Speed Adaptation via vehicle-to-infrastructure Communications. *IEEE Trans. Control Syst. Technol.* **2013**, *21*, 780–790. [CrossRef]
34. Tejado, I.; HosseinNia, S.H.; Vinagre, B.M.; Chen, Y.Q. Efficient control of a SmartWheel via Internet with compensation of variable delays. *Mechatronics* **2013**, *23*, 821–827. [CrossRef]
35. Tejado, I.; HosseinNia, S.H.; Vinagre, B.M. Adaptive gain-order fractional control for network-based applications. *Fract. Calc. Appl. Anal.* **2014**, *17*, 462–482. [CrossRef]
36. Pan, I.; Das, S. *Intelligent Fractional Order Systems and Control*; Chapter Gain and Order Scheduling for Fractional Order Controllers; Springer: Berlin/Heidelberg, Germany, 2013; pp. 147–157.
37. Dastjerdi, A.A.; Vinagre, B.M.; Chen, Y.Q.; HosseinNia, S.H. Linear fractional order controllers; A survey in the frequency domain. *Ann. Rev. Control* **2019**. [CrossRef]
38. Birs, I.; Muresan, C.; Nascu, I.; Ionescu, C. A Survey of Recent Advances in Fractional Order Control for Time Delay Systems. *IEEE Access* **2019**, *7*, 30951–30965. [CrossRef]

39. Petrás, I. *Handbook of Fractional Calculus with Applications*; Chapter Modified Versions of the Fractional-Order PID Controller; De Gruyter: Berlin, Germany, 2019; Volume 6, pp. 57–72.

40. Dastjerdi, A.A.; Saikumar, N.; HosseinNia, S.H. Tuning guidelines for fractional order PID controllers: Rules of thumb. *Mechatronics* **2018**, *56*, 26–36. [CrossRef]

41. Tepljakov, A.; Alagoz, B.B.; Yeroglu, C.; Gonzalez, E.; HosseinNia, S.H.; Petlenkov, E. FOPID Controllers and Their Industrial Applications: A Survey of Recent Results. In Proceedings of the 3rd IFAC Conference on Advances in Proportional-Integral-Derivative Control, Ghent, Belgium, 9–11 May 2018; Volume 51, pp. 25–30.

42. Shah, P.; Agashe, S. Review of fractional PID controller. *Mechatronics* **2016**, *38*, 29–41. [CrossRef]

43. Apte, A.; Thakar, U.; Joshi, V. Disturbance Observer Based Speed Control of PMSM Using Fractional Order PI Controller. *IEEE-CAA J. Autom. Sin.* **2019**, *6*, 316–326. [CrossRef]

44. AbouOmar, M.S.; Zhang, H.J.; Su, Y.X. Fractional Order Fuzzy PID Control of Automotive PEM Fuel Cell Air Feed System Using Neural Network Optimization Algorithm. *Energies* **2019**, *12*, 1435. [CrossRef]

45. Copot, D.; Ghita, M.; Ionescu, C.M. Simple Alternatives to PID-Type Control for Processes with Variable Time-Delay. *Processes* **2019**, *7*, 146. [CrossRef]

46. Deng, Y. Fractional-order fuzzy adaptive controller design for uncertain robotic manipulators. *Int. J. Adv. Robot. Syst.* **2019**, *16*, 1–10. [CrossRef]

47. Feliu-Talegon, D.; Feliu-Batlle, V.; Tejado, I.; Vinagre, B.M.; HosseinNia, S.H. Stable force control and contact transition of a single link flexible robot using a fractional-order controller. *ISA Trans.* **2019**, *89*, 139–157. [CrossRef] [PubMed]

48. Ren, H.P.; Fan, J.T.; Kaynak, O. Optimal Design of a Fractional-Order Proportional-Integer-Differential Controller for a Pneumatic Position Servo System. *IEEE Trans. Ind. Electr.* **2019**, *66*, 6220–6229. [CrossRef]

49. Zhang, F.; Yang, C.; Zhou, X.; Gui, W. Optimal Setting and Control Strategy for Industrial Process Based on Discrete-Time Fractional-Order (PID mu)-D-lambda. *IEEE Access* **2019**, *7*, 47747–47761. [CrossRef]

50. Mystkowski, A.; Kierdelewicz, A. Fractional-order water level control based on PLC: hardware-in-the-loop simulation and experimental validation. *Energies* **2018**, *11*, 2928. [CrossRef]

51. Podlubny, I. Geometric and Physical Interpretation of Fractional Integration and Fractional Differentiation. *Fract. Calc. Appl. Anal.* **2004**, *5*, 367–386.

52. Tavassoli, M.H.; Tavassoli, A.; Rahimi, M.R.O. The geometric and physical interpretation of fractional order derivatives of polynomial functions. *Differ. Geom. Dyn. Syst.* **2013**, *15*, 93–104.

53. Podlubny, I. *Fractional Differential Equations. An Introduction to Fractional Derivatives, Fractional Differential Equations, to Methods of Their Solution and Some of Their Applications*; Mathematics in Science and Engineering; Academic Press: Cambridge, MA, USA, 1999; Volume 198.

54. Hägglund, T. A unified discussion on signal filtering in PID control. *Control Eng. Pract.* **2013**, *21*, 994–1006. [CrossRef]

55. Isaksson, A.; Graebe, S. Derivative filter is an integral part of PID design. *IEE Proc. Control Theor. Appl.* **2002**, *149*, 41–45. [CrossRef]

56. Fractional Taylor Series for Caputo Fractional Derivatives. Construction of Numerical Schemes. Available online: http://www.fdi.ucm.es/profesor/lvazquez/calcfrac/docs/paper_Usero.pdf (accessed on 22 April 2019).

Σ *mathematics*

MDPI

Article

Audio Signal Processing Using Fractional Linear Prediction

Tomas Skovranek [1,*]**, Vladimir Despotovic** [2]

[1]   BERG Faculty, Technical University of Kosice, Nemcovej 3, 04200 Kosice, Slovakia
[2]   Technical Faculty in Bor, University of Belgrade, Vojske Jugoslavije 12, 19210 Bor, Serbia
*   Correspondence: tomas.skovranek@tuke.sk; Tel.: +421-55-602-5143

Received: 29 April 2019; Accepted: 24 June 2019; Published: 29 June 2019

Abstract: Fractional linear prediction (FLP), as a generalization of conventional linear prediction (LP), was recently successfully applied in different fields of research and engineering, such as biomedical signal processing, speech modeling and image processing. The FLP model has a similar design as the conventional LP model, i.e., it uses a linear combination of "fractional terms" with different orders of fractional derivative. Assuming only one "fractional term" and using limited number of previous samples for prediction, FLP model with "restricted memory" is presented in this paper and the closed-form expressions for calculation of FLP coefficients are derived. This FLP model is fully comparable with the widely used low-order LP, as it uses the same number of previous samples, but less predictor coefficients, making it more efficient. Two different datasets, MIDI Aligned Piano Sounds (MAPS) and Orchset, were used for the experiments. Triads representing the chords composed of three randomly chosen notes and usual Western musical chords (both of them from MAPS dataset) served as the test signals, while the piano recordings from MAPS dataset and orchestra recordings from the Orchset dataset served as the musical signal. The results show enhancement of FLP over LP in terms of model complexity, whereas the performance is comparable.

Keywords: audio signal processing; linear prediction; fractional derivative; musical signal

---

1. Introduction

The sinusoidal model is widely used for representation of pseudo-stationary signals, especially in audio coding [1] and musical signal processing [2]. Parameters of the sinusoidal model are determined frame-wise from the input audio/musical signal, and a sound is synthesized using the extracted parameters [3]. A pure tone can be represented as a single sine wave, whereas the musical chords are produced by combining three or more sine waves with different frequencies. In fact, any musical tone can be described as a combination of sine waves or its partials, each with its own amplitude, phase and frequency of vibration [4]. A sine wave can be fully described using three parameters: amplitude, phase and frequency. Obviously, such signal is redundant; hence, there is no need to encode and transmit each signal sample.

Linear prediction (LP) can be used to remove redundancy by predicting the current signal sample from the signal history, as the weighted linear combination of past samples. In that case, only the coefficients of the predictor need to be transmitted, not the signal samples themselves. While LP is extensively used for modeling speech signal [5–7], it did not prove to be the best choice for modeling audio signals. This is unexpected, since a signal represented by a combination of sine waves should be perfectly predicted using an LP model with an order twice larger than the number of sinusoids. The problem might be the fact that LP can model well signals with equally distributed tonal components in the Nyquist interval, which is not the case with audio, where tonal components are concentrated in a substantially smaller frequency region in comparison to the signal bandwidth [8]. This happens due to the fact that audio signals are usually sampled at a much higher frequency than the frequency of their tonal components. Nevertheless, there are applications of LP in audio coding algorithms using the so-called frequency-warped LP [9,10], where the unit delays are replaced by the first-order all-pass filter elements to adjust the frequency resolution in the spectral estimate to closely approximate the frequency resolution of human hearing [9]. LP is also used in acoustic echo cancelation [11], music dereverberation [12], audio signal classification [13] and audio/music onset detection [14,15].

The idea of using the signal history is fundamentally rooted in fractional calculus [16]. Fractional linear prediction (FLP), as a generalization of LP for fractional (arbitrary real) order derivatives, was recently used in electroencephalogram (EEG) [16,17] and electrocardiogram (ECG) signal modeling [18], as well as in speech coding [16,19–21]. While in [17–19] the full signal history is used for predicting the current signal sample, which is impractical from the implementation point of view, a model with restricted signal memory that uses only the recent signal samples and its applications is proposed in [21,22]. However, to the best of our knowledge, there are no applications of FLP in audio/musical signal processing. In this paper, we present FLP with memory restricted to maximum of four previous samples and apply it to prediction of randomly generated test chords, usual chords in Western music and piano parts extracted from the MIDI Aligned Piano Sounds dataset; and musical parts extracted from symphonies, ballets and other classical musical forms, and interpreted by symphonic orchestras, from the Orchset dataset.

The paper is organized as follows. Section 2 presents an overview of conventional LP and the FLP with "restricted memory". Datasets used for experiments are described in Section 3. The numerical results using the test chords, piano and orchestra musical parts are discussed in Section 4, followed by concluding remarks in Section 5.

2. Linear Prediction

2.1. Conventional Linear Prediction

Let the signal $x(t)$ represent a linear and stationary stochastic process, where $x_{[n]} = x(nT)$ is the nth signal sample at arbitrary time t, and T is the sampling period. The signal $x(t)$ at time instance $t = nT$ is modeled as the linear combination of p previous signal samples:

$$\hat{x}_{[n]} = \sum_{i=1}^{p} a_i x_{[n-i]},$$ (1)

where $\hat{x}_{[n]}$ denotes the predicted signal sample and a_i are the linear predictor coefficients. The order of a linear predictor denotes the number of linear predictor coefficients, which is equal to the number of samples used for prediction.

The prediction error $e_{[n]} = x_{[n]} - \hat{x}_{[n]}$ is defined as the deviation of the predicted signal $\hat{x}$ from the original signal x, and the mean-squared prediction error is equal to:

$$J = E\left[e_{[n]}^2\right] = E\left[x_{[n]} - \sum_{i=1}^{p} a_i x_{[n-i]}\right]^2,$$ (2)

where $E[\cdot]$ is the mathematical expectation. The optimal predictor coefficients a_i can be determined by equating the first derivative of J, with respect to a_i, to zero. After some manipulation, we obtain:

$$\sum_{i=1}^{p} a_i R_{xx}(k-i) = R_{xx}(k), \quad k = 1, 2, \ldots, p,$$ (3)

where $R_{xx}(k) = E\left[x_{[n]} x_{[n-k]}\right]$ denotes the autocorrelation function at lag k. Equation (3) is known as the Yule–Walker equation [7] and can be rewritten in the matrix form as:

$$\mathbf{R}_{xx} \cdot \mathbf{a} = \mathbf{r}_{xx}, \tag{4}$$

where

$$\mathbf{R}_{xx} = \begin{bmatrix} R_{xx}(0) & R_{xx}(1) & R_{xx}(2) & \cdots & R_{xx}(p-1) \\ R_{xx}(1) & R_{xx}(2) & R_{xx}(3) & \cdots & R_{xx}(p-2) \\ \vdots & \vdots & \vdots & \ddots & \vdots \\ R_{xx}(p-1) & R_{xx}(p-2) & R_{xx}(p-3) & \cdots & 0 \end{bmatrix},$$

$$\mathbf{a} = \begin{bmatrix} a_1 & a_2 & a_3 & \cdots & a_p \end{bmatrix}^T,$$

$$\mathbf{r}_{xx} = \begin{bmatrix} R_{xx}(1) & R_{xx}(2) & R_{xx}(3) & \cdots & R_{xx}(p) \end{bmatrix}^T.$$

The optimal linear predictor coefficients **a** can be found from:

$$\mathbf{a} = \mathbf{R}_{xx}^{-1} \cdot \mathbf{r}_{xx}. \tag{5}$$

2.2. Fractional Linear Prediction with "Restricted Memory"

FLP is a generalization of LP using the fractional-order derivatives. Using the analogy from LP, the nth signal sample can be represented as the linear combination of q "fractional terms", and can be written as [16]:

$$\hat{x}_{[n]} = \sum_{i=1}^{q} a_i D^{\alpha_i} x_{[n-1]}, \tag{6}$$

where $\hat{x}_{[n]}$ is the estimate of the nth signal sample, q is the number of "fractional terms" used for the prediction, a_i are the FLP coefficients, and $D^{\alpha} x_{[n-1]}$ are the fractional derivatives of order α_i of the time-delayed signal, where $\alpha_i \in \mathbb{R}$.

The fractional derivative D^{α} can be approximated by the Grünwald–Letnikov (GL) definition of a function $x(t)$ at time instant t [23]:

$$_a D_t^{\alpha} x(t) = \lim_{h \to 0} \frac{1}{h^{\alpha}} \sum_{j=0}^{\lfloor \frac{t-a}{h} \rfloor} (-1)^j \binom{\alpha}{j} x(t - jh), \tag{7}$$

where h is the sampling period, a and t are lower and upper limits of differentiation, and $\alpha \in \mathbb{R}$ is the order of fractional differentiation. Note that the upper limit of summation tends to infinity. Accounting only for the recent history of the signal, i.e., replacing the lower limit a by the the moving lower limit $t - L$ (L is the memory length), the "short memory" principle [23] is employed. Due to this approximation, the number of addends in Equation (7) is not greater than $K = \lfloor L/h \rfloor$. For $t = nh$, Equation (7) becomes:

$$D^{\alpha} x(nh) = \lim_{h \to 0} \frac{1}{h^{\alpha}} \sum_{j=0}^{K} (-1)^j \binom{\alpha}{j} x((n - j)h). \tag{8}$$

Replacing $x(nh)$ with $x_{[n]}$, and assuming that in the signal prediction only the past samples are used for the estimation of the predicted signal sample, without including the current sample, i.e., introducing a time-delay in Equation (8) of one sample, one gets:

$$D^{\alpha} x_{[n-1]} = h^{-\alpha} \sum_{j=0}^{K} (-1)^j \binom{\alpha}{j} x_{[n-1-j]}. \tag{9}$$

Taking into account only one "fractional term" from Equation (6), i.e., when $q = 1$, one obtains [21,22]:

$$\hat{x}_{[n]} = a D^{\alpha} x_{[n-1]}. \tag{10}$$

Considering $K \in \mathbb{I}$ as the upper limit of the summation in Equation (9), i.e., for $K = 1$:

$$D^\alpha x_{[n-1]} = \frac{1}{h^\alpha}\left(x_{[n-1]} - \alpha x_{[n-2]}\right),\tag{11}$$

$K = 2$:

$$D^\alpha x_{[n-1]} = \frac{1}{h^\alpha}\left(x_{[n-1]} - \alpha x_{[n-2]} - \frac{\alpha(1-\alpha)}{2}x_{[n-3]}\right),\tag{12}$$

and $K = 3$:

$$D^\alpha x_{[n-1]} = \frac{1}{h^\alpha}\left(x_{[n-1]} - \alpha x_{[n-2]} - \frac{\alpha(1-\alpha)}{2}\left(x_{[n-3]} + \frac{2-\alpha}{3}x_{[n-4]}\right)\right),\tag{13}$$

we get three modifications of FLP model with "restricted memory" (Equation (10)), which use the memory (M) of two, three, and four samples, respectively.

Employing the memory of two samples, i.e., substituting $D^\alpha x_{[n-1]}$ from Equation (11) into Equation (10), the two-sample FLP model is defined as:

$$\hat{x}_{[n]} = \frac{a}{h^\alpha}\left(x_{[n-1]} - \alpha x_{[n-2]}\right),\tag{14}$$

and the prediction error is evaluated as $e_{[n]} = x_{[n]} - \hat{x}_{[n]}$. Minimizing the mean squared prediction error $J = E\left[e_{[n]}^2\right]$ and substituting the autocorrelation function, the optimal coefficient a can be found. After some manipulation, the optimal FLP parameter can be written as:

$$a = h^\alpha \frac{R_{xx}(1) - \alpha R_{xx}(2)}{R_{xx}(0) - 2\alpha R_{xx}(1) + \alpha^2 R_{xx}(0)}.\tag{15}$$

In case the order of fractional derivative α tends to zero, we get:

$$\lim_{\alpha\to 0} a = \lim_{\alpha\to 0} h^\alpha \frac{R_{xx}(1) - \alpha R_{xx}(2)}{R_{xx}(0) - 2\alpha R_{xx}(1) + \alpha^2 R_{xx}(0)} = \frac{R_{xx}(1)}{R_{xx}(0)},\tag{16}$$

i.e., the optimal first-order linear predictor is only a special case of the proposed FLP model with "restricted memory" using the memory of two previous samples.

Considering the FLP model with "restricted memory" of three samples, where $D^\alpha x_{[n-1]}$ is estimated using Equation (12), the predicted sample becomes:

$$\hat{x}_{[n]} = \frac{a}{h^\alpha}\left(x_{[n-1]} - \alpha x_{[n-2]} - \frac{\alpha(1-\alpha)}{2}x_{[n-3]}\right).\tag{17}$$

Minimizing the mean squared prediction error $J = E\left[e_{[n]}^2\right]$, the optimal coefficient a can be found as:

$$a = h^\alpha \frac{R_{xx}(1) - \alpha R_{xx}(2) - \frac{\alpha(1-\alpha)}{2}R_{xx}(3)}{R_{xx}(0) - 2\alpha\left(R_{xx}(1) - \frac{\alpha-1}{2}R_{xx}(2)\right) + \alpha^2\left(R_{xx}(0) - (\alpha-1)R_{xx}(1) + \frac{(\alpha-1)^2}{4}R_{xx}(0)\right)}.\tag{18}$$

As in the case of FLP model with two-sample memory, when the order of fractional derivative α tends to zero, the computation of the FLP coefficient a reduces to $a = R_{xx}(1)/R_{xx}(0)$, meaning that the first-order LP is a special case of the FLP model with "restricted memory" using the memory of three previous samples.

The last modification of the presented FLP model with "restricted memory" (Equation (10)) is taking into account the memory of four previous samples, i.e., $D^\alpha x_{[n-1]}$ is estimated using Equation (13):

$$\hat{x}_{[n]} = \frac{a}{h^\alpha}\left(x_{[n-1]} - \alpha x_{[n-2]} - \frac{\alpha(1-\alpha)}{2}\left(x_{[n-3]} + \frac{2-\alpha}{3}x_{[n-4]}\right)\right). \tag{19}$$

Computing the prediction error $e_{[n]} = x_{[n]} - \hat{x}_{[n]}$ and minimizing the mean squared prediction error $J = E\left[e_{[n]}^2\right]$ by finding the first derivative of J with respect to a and equating to zero, optimal coefficient a is obtained in the form:

$$a = h^\alpha\frac{R_{xx}(1) - \alpha R_{xx}(2) - \frac{\alpha(1-\alpha)}{2}\left(R_{xx}(3) - \frac{\alpha-2}{3}R_{xx}(4)\right)}{R_{xx}(0) - 2\alpha R_{xx}(1) + \alpha^2 R_{xx}(0) + \frac{\alpha^2(\alpha-1)^2}{4}\#1 + \alpha(\alpha-1)\#2} \tag{20}$$

where

$$\#1 = \left(R_{xx}(0) - \frac{2\alpha-4}{3}R_{xx}(1) + \frac{(\alpha-2)^2}{9}R_{xx}(0)\right),$$

$$\#2 = \left(R_{xx}(2) - \alpha R_{xx}(1) - \frac{\alpha-2}{3}R_{xx}(3) + \frac{\alpha(\alpha-2)}{3}R_{xx}(2)\right).$$

Again, as in the case of FLP model with two-sample and three-sample memory, in the case of using the memory of four samples, when the order of fractional derivative α tends to zero, the computation of the FLP coefficient a is reduced to $a = R_{xx}(1)/R_{xx}(0)$. This confirms that the proposed FLP models with the "restricted memory" are generalizations of the low-order LP, i.e., the first-order LP is only a special case of the presented FLP model.

It was proven in [21,22] that the parameter α of the FLP model with "restricted memory" can be estimated as the inverse of the number of samples used by the FLP model, i.e., $\alpha = 1/M$. Thus, the order of fractional differentiation is in this paper assumed fixed, with the values $\alpha = 0.5$ for FLP model with two-sample memory, $\alpha = 0.33$ for FLP model with three-sample memory, and $\alpha = 0.25$ for FLP model with four-sample memory. It follows that the FLP model with "restricted memory" practically uses only one predictor coefficient, which has to be encoded and transmitted, regardless of the number of previous samples used for prediction.

3. Datasets

3.1. MAPS Dataset

The MIDI Aligned Piano Sounds (MAPS) dataset contains 65 h of stereo audio recordings sampled at 44.1 kHz with 16 bit resolution (CD quality), recorded either using the software-based sound generation, or the Disklavier piano [24,25]. The dataset contains four subsets: isolated notes (ISOL); chords composed of randomly chosen notes (RAND); usual chords in Western music (UCHO); and piano classical music pieces (MUS). The audio samples were recorded in different recording conditions (e.g., studio, jazz club, church, and concert hall). RAND, UCHO and MUS subsets were used in the experiments using all four recording conditions.

3.2. Orchset Dataset

Orchset database contains 64 mono and stereo audio recordings, sampled at 44.1 kHz, extracted from symphonies, ballets and other classical musical forms and interpreted by symphonic orchestras [26]. The lengths of the recordings are 10–32 s (mean 22.1 s, standard deviation 6.1 s), the number of recordings per composer is 1–13, with 15 composers in total. Music excerpts were selected to have a dominant melody, maximizing the existence of voiced segments per excerpt. In all excerpts, the melody was played using more than one instrument from the instrument section, except for one excerpt where only oboe was used (with orchestral accompaniment).

3.3. Signal Preprocessing

In signal processing applications, e.g., when processing speech or audio signal that are non-stationary signals, the signal is usually divided into short-time windows, denoted as frames, where the signal is approximately stationary. In the case of audio signal, the frame length is typically 10–120 ms [27,28]. In this study, the experiments were performed using three different frame-sizes, equal to 10 ms, 60 ms and 120 ms.

The audio signal may contain silent periods, usually at the beginning or at the end of a signal. This was especially evident in RAND and UCHO subsets of the MAPS dataset, where the silence periods were even longer than the signal itself. Modeling silent frames is unnecessary since the resources are spent on parts of the signal which do not contribute to signal reconstruction. Therefore, the silence frames were removed before further processing. Furthermore, DC offset was removed from the audio signal, as the signal compression, or any other processing of the signal that includes the absolute signal levels may lead to distortions and other non-desirable results. Finally, all stereo recordings were converted to mono by combining left and right channels prior to further processing.

4. Numerical Results and Discussion

The proposed FLP with "restricted memory" given in Equation (10) with the memory of two (Equation (14)), three (Equation (17)) and four samples (Equation (19)) was compared to conventional low-order LP using the same signal history. Experiments were performed using two test signals: the three-note chords composed of randomly chosen notes (MAPS–RAND subset), usual three-notes Western musical chords (MAPS–UCHO subset), and two musical signals: piano recordings (MAPS–MUS subset) and orchestra recordings (Orchset). The signals belonging to one recording condition (studio, jazz club, church, or concert hall) of the particular dataset were concatenated to one signal prior to applying either LP or FLP.

The prediction gain (PG) served as the predictor performance measure, defined as the ratio between the variance of the input signal and the variance of the prediction error measured in decibels:

$$PG\ (dB) = 10\log_{10}\frac{\sigma_x^2}{\sigma_{e_p}^2}. \tag{21}$$

The smaller is the error generated by the predictor, the higher is the gain [29].

Experiments

The results for the randomly generated chords (MAPS–RAND subset) for different recording conditions (studio, jazz club, church, and concert hall) using four low-order LP models (first-order, second-order, third-order and fourth-order) and FLP models with the two-sample, three-sample and four-sample memory are presented in Table 1. The results show that the first-order LP is inappropriate; however, increasing the prediction-order beyond the second-order LP is not necessary, as it does not bring significant improvement. Similar behavior can be observed for FLP models, where the best performing model is the one with the two-sample memory. For the frames having 120 ms length, its performance is only slightly lower than the performance of the second-order LP, albeit obtained using only one predictor coefficient (note that the second-order LP that also uses the memory of two samples, requires the optimization of two predictor coefficients). By decreasing the frame length, the performance of both LP and FLP decrease, but with FLP approaching LP for the memory of three and four samples. Note that the results for FLP with the memory of three and four samples were obtained using two and three predictor coefficients less than in the case of the third-order and fourth-order LP.

The prediction results for the chords composed of three randomly chosen notes from the MAPS–RAND subset are also presented in Figure 1, where the prediction error using the second-order, third-order and fourth-order LP (black solid line) is compared to the prediction error obtained using the FLP model with two-sample, three-sample and four-sample memory (red dot-dashed line). Ten characteristic frames with the length of 60 ms are shown in the figure. The results confirm that the performance of the second-order LP and the FLP with two-sample memory is comparable for the signals recorded under different conditions (studio, jazz club, church, and concert hall), and the difference between the prediction error of the LP and FLP models is generally increasing with the length of the used memory.

Table 1. Prediction gain (dB) for the chords composed of three randomly chosen notes (MAPS–RAND subset).

			MAPS–RAND			
			Studio	Jazz	Church	Concert
120 ms	LP	First-order	17.41	17.53	19.10	15.48
		Second-order	23.94	23.91	26.25	22.51
		Third-order	24.85	24.52	26.89	23.55
		Fourth-order	25.25	24.79	27.15	23.96
	FLP	Two-sample memory	23.40	23.36	25.82	22.14
		Three-sample memory	23.41	23.68	26.02	22.01
		Four-sample memory	23.11	25.06	25.88	21.63
60 ms	LP	First-order	17.15	17.35	18.90	15.23
		Second-order	22.90	22.82	25.15	21.43
		Third-order	23.51	23.42	25.70	22.14
		Fourth-order	23.85	23.66	25.93	22.51
	FLP	Two-sample memory	22.32	22.25	24.71	21.07
		Three-sample memory	22.47	22.66	25.01	21.08
		Four-sample memory	22.28	22.73	24.96	20.81
10 ms	LP	First-order	16.35	16.58	18.13	14.65
		Second-order	19.82	19.95	21.65	18.86
		Third-order	20.28	20.48	22.19	19.29
		Fourth-order	20.46	20.68	22.38	19.50
	FLP	Two-sample memory	19.30	19.37	21.22	18.50
		Three-sample memory	19.74	19.96	21.78	18.80
		Four-sample memory	19.81	20.17	21.94	18.76

Similar behavior as in case of randomly generated chords can be observed when using usual three-notes Western musical chords (MAPS–UCHO subset). Again, the performance of FLP with two-sample memory is comparable to the second-order LP for all frames, although FLP is using one coefficient less (see Table 2).

Ten characteristic frames with the length of 60 ms are shown in Figure 2 for the MAPS–UCHO subset, where the prediction error using the second-order, third-order and fourth-order LP (black solid line) is compared to the prediction error obtained using the FLP model with two-sample, three-sample and four-sample memory (red dot-dashed line). The results confirm that the performance of the second-order LP and the FLP with two-sample memory is comparable for the signals recorded under different conditions (studio, jazz club, church, and concert hall), and also that the difference between the prediction errors of the LP and FLP models is increasing with the length of the used memory.

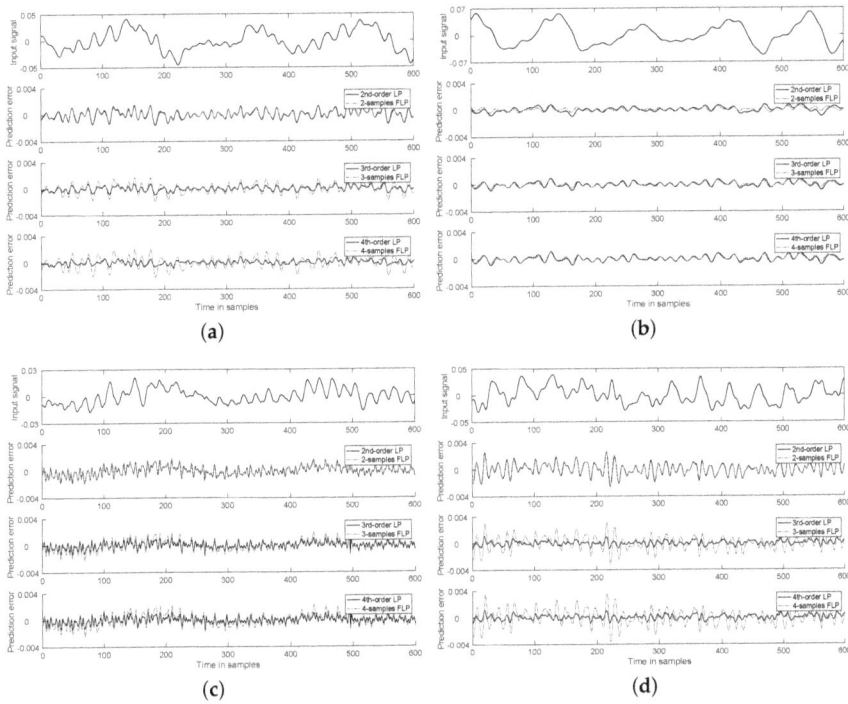

Figure 1. The prediction error results for the random chords (MAPS–RAND) for second-order, third-order and fourth-order LP and the FLP with the two-sample, three-sample, and four-sample memory: (**a**) studio recording; (**b**) jazz club recording; (**c**) church recording; and (**d**) concert hall recording.

Table 2. Prediction gain (dB) for the usual Western music three-notes chords (MAPS–UCHO subset).

			MAPS–UCHO			
			Studio	Jazz	Church	Concert
120 ms	LP	First-order	17.03	18.54	18.74	17.44
		Second-order	24.51	25.75	26.62	25.22
		Third-order	25.25	26.29	27.12	26.02
		Fourth-order	25.61	26.52	27.34	26.39
	FLP	Two-sample memory	23.95	25.08	26.29	24.92
		Three-sample memory	23.90	25.44	26.46	24.78
		Four-sample memory	23.57	25.47	26.29	24.37
60 ms	LP	First-order	16.83	18.37	18.53	17.15
		Second-order	23.53	24.58	25.42	24.07
		Third-order	24.04	25.10	25.91	24.62
		Fourth-order	24.32	25.29	26.08	24.93
	FLP	Two-sample memory	22.97	23.89	25.07	23.76
		Three-sample memory	23.05	24.36	25.36	23.76
		Four-sample memory	22.82	24.47	25.30	23.46
10 ms	LP	First-order	16.07	17.46	17.71	16.38
		Second-order	20.44	21.15	21.77	20.85
		Third-order	20.87	21.74	22.32	21.29
		Fourth-order	21.01	21.93	22.49	21.45
	FLP	Two-sample memory	19.95	20.51	21.45	20.57
		Three-sample memory	20.34	21.15	21.99	20.90
		Four-sample memory	20.37	21.42	22.14	20.88

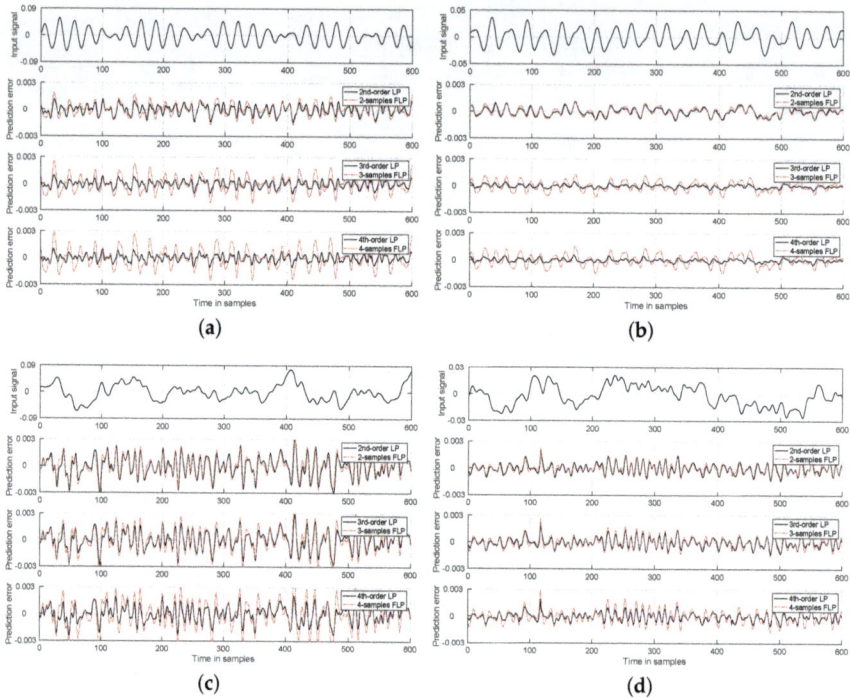

Figure 2. The prediction error results for the three-notes chords (MAPS–UCHO) for second-order, third-order and fourth-order LP and the FLP with the two-sample, three-sample, and four-sample memory: (**a**) studio recording; (**b**) jazz club recording; (**c**) church recording; and (**d**) concert hall recording.

The results for the piano music excerpts using MAPS–MUS subset are also presented for three different frame sizes, i.e., 10 ms, 60 ms and 120 ms (see Table 3). For shorter frames (10 ms), the performance of FLP is always comparable to the performance of the corresponding LP that uses the same signal memory. For longer frames, PG of FLP is comparable to PG of the corresponding LP for jazz club and church recording conditions, while the performance deteriorates by 1–2 dB only for FLP with the memory of three and four samples for studio and concert recording conditions, suggesting that FLP is better suited for signals recorded in reverberant or non-ideal acoustical conditions. Note that FLP always uses only one predictor coefficient, regardless of the signal memory used for prediction. For example, for the FLP with the four-sample memory, comparable performance is obtained to the corresponding fourth-order LP, but with three predictor coefficients less that need to be optimized. This can lead to substantial savings in bit rate, as predictor coefficients need to be encoded and transferred to receiver end. Furthermore, note that better performance is obtained using longer frames for both LP and FLP; hence, more frequent coefficient update does not bring any improvement.

The last experiment was performed using the orchestra music excerpts from the Orchset dataset. Since LP models are, in general, known to perform well on piano music, we tested the performance of our model on a more challenging music signal played by the orchestra (see Table 3). The performance of FLP in comparison to LP is lower than in piano music; however, the model with two-sample memory is still comparable to the corresponding second-order LP for all frame lengths. Third- and fourth-order LP models perform better than FLP at the expense of two and three additional coefficients, respectively.

Table 3. Prediction gain (dB) for musical signal of classical music pieces played by piano (MAPS–MUS subset) and the classical music pieces performed by orchestra (Orchset dataset).

			MAPS–MUS				Orchset
			Studio	Jazz	Church	Concert	
120 ms	LP	First-order	20.54	22.13	21.90	19.60	18.12
		Second-order	31.60	34.04	32.95	30.21	26.82
		Third-order	32.36	34.52	33.51	31.24	27.94
		Fourth-order	32.86	34.75	33.78	31.74	28.15
	FLP	Two-sample memory	31.59	34.02	32.94	30.18	26.70
		Three-sample memory	31.20	34.25	32.98	29.69	26.03
		Four-sample memory	30.55	33.98	32.65	28.96	25.29
60 ms	LP	First-order	20.49	22.00	21.79	19.58	18.08
		Second-order	30.27	32.05	31.28	29.10	26.18
		Third-order	30.81	32.63	31.87	29.82	26.99
		Fourth-order	31.17	32.80	32.07	30.22	27.18
	FLP	Two-sample memory	30.25	32.04	31.26	29.08	26.09
		Three-sample memory	30.14	32.56	31.57	28.83	25.56
		Four-sample memory	29.66	32.44	31.39	28.25	24.91
10 ms	LP	First-order	19.68	20.94	20.77	18.90	17.53
		Second-order	25.18	25.92	25.66	24.60	23.01
		Third-order	25.75	26.75	26.40	25.15	23.37
		Fourth-order	25.92	27.01	26.62	25.35	23.49
	FLP	Two-sample memory	25.17	25.92	25.66	24.57	23.00
		Three-sample memory	25.70	26.74	26.39	24.97	22.93
		Four-sample memory	25.64	27.00	26.52	24.81	22.62

When evaluating the prediction error in case of using musical signals from the MAPS–MUS subset (see Figure 3) under the same recording conditions as in previous experiments (e.g., studio, jazz club, church, and concert hall), an interesting observation can be made, i.e., the difference between the prediction error of the LP and FLP models is not increasing that significantly with the length of the used memory (especially for the jazz club and church recording conditions), as was the case of using signals representing chords. Furthermore, it is obvious that the second-order LP and the FLP with two-sample memory for the shown signals perform at the same level for all four recording conditions. Similar behavior is present in the case of using orchestra music excerpts from the Orchset dataset (see Figure 4). Please note that, in Figures 3 and 4, again ten characteristic frames with the length of 60 ms are shown, and that the prediction error using the second-order, third-order and fourth-order LP (black solid line) is compared to the prediction error obtained using the FLP model with two-sample, three-sample and four-sample memory (red dot-dashed line).

Here, it should be emphasized that LP and FLP models always use the same number of previous samples (two, three and four) that allows a fair comparison. Furthermore, it is important to emphasize that all FLP models show comparable performance in comparison to LP models, even though they use only two coefficients, i.e., one predictor coefficient a and one order of fractional derivative α, in comparison to LP models that use two, three and four predictor coefficients (based on the order of the LP predictor). Moreover, the order of fractional differentiation α does not have to be computed or optimized. It might be estimated as the inverse of the predictor memory, as previously shown in [21,22], resulting in only one FLP coefficient that has to be encoded and transmitted. This makes the proposed FLP significantly more efficient than LP.

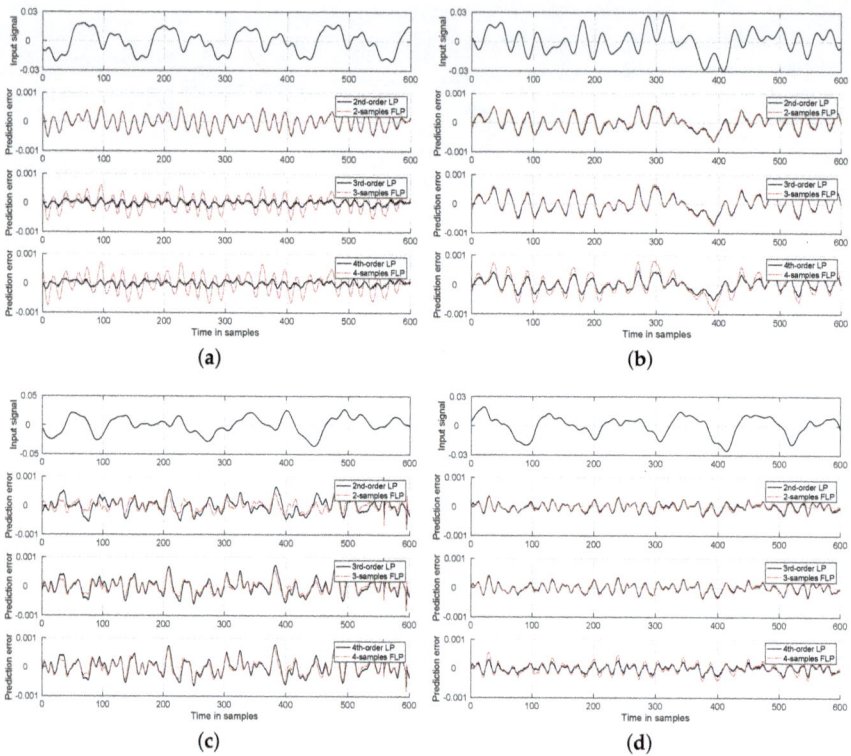

Figure 3. The prediction error results for the musical signals (MAPS–MUS) for second-order, third-order and fourth-order LP and the FLP with the two-sample, three-sample, and four-sample memory: (**a**) studio recording; (**b**) jazz club recording; (**c**) church recording; and (**d**) concert hall recording.

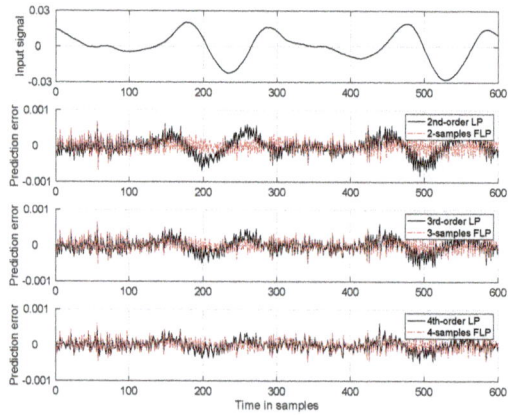

Figure 4. The prediction error results for the musical signal (Strauss–BlueDanube–ex1, from the Orchset) for second-order, third-order and fourth-order LP and the FLP with the two-sample, three-sample, and four-sample memory.

5. Conclusions

Fractional linear prediction with "restricted memory" that uses two, three, and four previous samples, respectively, for audio signal prediction is discussed in this work and the closed-form expressions for the FLP predictor coefficient are derived. Two datasets were used for the experiments to test the performance of the model and compare it to linear prediction, i.e., MAPS dataset, which contains chords composed of randomly chosen notes, usual chords in Western music, and piano music excerpts; and Orchset dataset, which contains music excerpts, extracted from symphonies, ballets and other classical musical forms, and interpreted by symphonic orchestras.

Using the same number of previous samples for prediction, the results show that FLP is better suited for prediction of audio signal than the conventional low-order LP models, since it provides comparable performance, even though it uses less parameters (one predictor coefficients and one order of fractional derivative). Furthermore, the order of fractional derivative does not have to be optimized and can be assumed as the inverse of the memory length of the FLP model, making it even more efficient in comparison to LP model, where the number of predictor coefficients is always equal to the predictor order. For example, FLP with the memory of four samples requires only one predictor coefficient, whereas the corresponding fourth-order LP requires four predictor coefficients, at similar performance. Therefore, substantial savings in transmission costs are possible.

Author Contributions: Investigation, T.S.; Methodology, T.S. and V.D.; Software, T.S.; Validation, V.D.; Visualization, V.D.; and Writing–original draft, T.S. and V.D.

Funding: This research was funded in part by the Slovak Research and Development Agency under Grants APVV-14-0892, SK-SRB-18-0011, and SK-AT-2017-0015; in part by the Slovak Grant Agency for Science under Grant VEGA 1/0365/19; in part by the Ministry of Education, Science and Technological Development of the Republic of Serbia under Grant 337-00-107/2019-09/11; and in part by the framework of the COST Action CA15225.

Conflicts of Interest: The authors declare no conflict of interest.

References

1. Purnhagen, H.; Meine, N. HILN–The MPEG-4 Parametric Audio Coding Tools. In Proceedings of the 2000 IEEE International Symposium on Circuits and Systems. Emerging Technologies for the 21st Century, Geneva, Switzerland, 28–31 May 2000; Volume 3, pp. 201–204.
2. Marchand, S.; Strandh, R. InSpect and ReSpect: Spectral Modeling, Analysis and Real-Time Synthesis Software Tools for Researchers and Composers. In Proceedings of the Int. Computer Music Conference (ICMC 1999), Beijing, China, 22–27 October 1999; pp. 341–344.
3. Lagrange, M.; Marchand, S. Long Interpolation of Audio Signals Using Linear Prediction in Sinusoidal Modeling. *J. Audio Eng. Soc.* **2005**, *53*, 891–905.
4. Thompson, W.F. *Music, Thought, and Feeling: Understanding the Psychology of Music*; Oxford University Press: Oxford, UK; New York, NY, USA, 2008.
5. Atal, B.S. The History of Linear Prediction. *IEEE Signal Process. Mag.* **2006**, *23*, 154–161. [CrossRef]
6. Benesty, J.; Chen, J.; Huang, Y. Linear Prediction. In *Springer Handbook of Speech Processing*; Benesty, J., Sondhi, M.M., Huang, Y., Eds.; Springer: Berlin, Germany, 2007; Chapter 7, pp. 121–133.
7. Vaidyanathan, P.P. *The Theory of Linear Prediction*; Synthesis Lectures on Signal Processing; Morgan & Claypool: San Rafael, CA, USA, 2008.
8. van Waterschoot, T.; Moonen, M. Comparison of linear prediction models for audio signals. *EURASIP J. Audio Speech Music Process.* **2009**, *2008*. [CrossRef]
9. Harma, A.; Laine, U.K. A comparison of warped and conventional linear predictive coding. *IEEE Trans. Speech Audio Process.* **2001**, *9*, 579–588. [CrossRef]
10. Deriche, M.; Ning, D. A novel audio coding scheme using warped linear prediction model and the discrete wavelet transform. *IEEE Trans. Audio Speech Lang. Process.* **2006**, *14*, 2039–2048. [CrossRef]
11. Van Waterschoot, T.; Rombouts, G.; Verhoeve, P.; Moonen, M. Double-talk-robust prediction error identification algorithms for acoustic echo cancellation. *IEEE Trans. Signal Process.* **2007**, *55*, 846–858. [CrossRef]

12. Mahkonen, K.; Eronen, A.; Virtanen, T.; Helander, E.; Popa, V.; Leppanen, J.; Curcio, I.D. Music dereverberation by spectral linear prediction in live recordings. In Proceedings of the 16th Int. Conference on Digital Audio Effects (DAFx-13), Maynooth, Ireland, 2–6 September 2013; pp. 1–4.

13. Grama, L.; Rusu, C. Audio signal classification using Linear Predictive Coding and Random Forests, Bucharest, Romania. In Proceedings of the International Conference on Speech Technology and Human-Computer Dialogue (SpeD 2017), Bucharest, Romania, 6–9 July 2017.

14. Glover, J.; Lazzarini, V.; Timoney, J. Real-time detection of musical onsets with linear prediction and sinusoidal modeling. *EURASIP J. Adv. Signal Process.* **2011**, *2011*, 68. [CrossRef]

15. Marchi, E.; Ferroni, G.; Eyben, F.; Gabrielli, L.; Squartini, S.; Schuller, B. Multi-resolution linear prediction based features for audio onset detection with bidirectional LSTM neural networks. In Proceedings of the IEEE International Conference on Acoustic, Speech and Signal Processing (ICASSP 2014), Florence, Italy, 4–9 May 2014; pp. 2183–2187.

16. Skovranek, T.; Despotovic, V. Signal Prediction using Fractional Derivative Models. In *Handbook of Fractional Calculus with Applications*; Baleanu, D., Lopes, A.M., Eds.; Walter de Gruyter GmbH: Berlin/Munich, Germany; Boston, MA, USA, 2019; Volume 8, Chapter 7, pp. 179–206.

17. Joshia, V.; Pachori, R.B.; Vijesh, A. Classification of ictal and seizure-free EEG signals using fractional linear prediction. *Biomed. Signal Process. Control* **2014**, *9*, 1–5. [CrossRef]

18. Talbi, M.L.; Ravier, P. Detection of PVC in ECG Signals Using Fractional Linear Prediction. *Biomed. Signal Process. Control* **2016**, *23*, 42–51. [CrossRef]

19. Assaleh, K.; Ahmad, W.M. Modeling of Speech Signals Using Fractional Calculus. In Proceedings of the 9th International Symposium on Signal Processing and Its Applications (ISSPA'07), Sharjah, UAE, 12–15 February 2007; pp. 1–4.

20. Despotovic, V.; Skovranek, T. Fractional-order Speech Prediction. In Proceedings of the International Conference on Fractional Differentiation and its Applications (ICFDA'16), Novi Sad, Serbia, 18–20 July 2016; pp. 124–127.

21. Despotovic, V.; Skovranek, T.; Peric, Z. One-parameter fractional linear prediction. *Comput. Electr. Eng. Spec. Issue Signal Process.* **2018**, *69*, 158–170. [CrossRef]

22. Skovranek, T.; Despotovic, V.; Peric, Z. Optimal Fractional Linear Prediction With Restricted Memory. *IEEE Signal Process. Lett.* **2019**, *26*, 760–764. [CrossRef]

23. Podlubny, I. *Fractional Differential Equations*; Academic Press: San Diego, CA, USA, 1999.

24. Emiya, V.; Badeau, R.; David, B. Multipitch estimation of piano sounds using a new probabilistic spectral smoothness principle. *IEEE Trans. Audio Speech Lang. Process.* **2010**, *18*, 1643–1654. [CrossRef]

25. Emiya, V. Transcription Automatique de la Musique de Piano. Ph.D. Thesis, Telecom ParisTech, Paris, France, October 2008.

26. Bosch, J.; Marxer, R.; Gomez, E. Evaluation and Combination of Pitch Estimation Methods for Melody Extraction in Symphonic Classical Music. *J. New Music Res.* **2016**, *45*, 101–117. [CrossRef]

27. Driedger, J.; Mueller, M. A Review of Time-Scale Modification of Music Signals. *Appl. Sci.* **2016**, *6*, 57. [CrossRef]

28. Theodoridis, S.; Koutroumbas, K. *Pattern Recognition*, 4th ed.; Academic Press: San Diego, CA, USA, 2008.

29. Chu, W.C. *Speech Coding Algorithms: Foundation and Evolution of Standardized Coders*; John Wiley & Sons: Hoboken, NJ, USA, 2003.

mathematics

MDPI

Article

Time-Fractional Diffusion-Wave Equation with Mass Absorption in a Sphere under Harmonic Impact

Bohdan Datsko [1,2], Igor Podlubny [3] and Yuriy Povstenko [4,*]

[1] Faculty of Mathematics and Applied Physics, Rzeszow University of Technology, Powstancow Warszawy 8, 35-959 Rzeszow, Poland; datskob@prz.edu.pl
[2] Institute for Applied Problems of Mechanics and Mathematics NAS of Ukraine, 79060 Lviv, Ukraine
[3] BERG Faculty, Technical University of Kosice, B. Nemcovej 3, 04200 Kosice, Slovakia; igor.podlubny@tuke.sk
[4] Faculty of Mathematical and Natural Sciences, Jan Dlugosz University in Czestochowa, Armii Krajowej 13/15, 42-200 Czestochowa, Poland
* Correspondence: j.povstenko@ajd.czest.pl; Tel.: +48-343-612-269

Received: 24 April 2019; Accepted: 12 May 2019; Published: 16 May 2019

Abstract: The time-fractional diffusion equation with mass absorption in a sphere is considered under harmonic impact on the surface of a sphere. The Caputo time-fractional derivative is used. The Laplace transform with respect to time and the finite sin-Fourier transform with respect to the spatial coordinate are employed. A graphical representation of the obtained analytical solution for different sets of the parameters including the order of fractional derivative is given.

Keywords: fractional calculus; mass absorption; diffusion-wave equation; Caputo derivative; harmonic impact; Laplace transform; Fourier transform; Mittag-Leffler function

---

1. Introduction

The classical parabolic diffusion equation with heat or mass absorption [1]

$$\frac{\partial u}{\partial t} = a\Delta u - bu \tag{1}$$

also describes bioheat transfer, lateral surface mass or heat exchange in a thin plate, heating of tissue during laser treatment irradiation, etc. (see, for example [2–5]). The Klein-Gordon equation

$$\frac{\partial^2 u}{\partial t^2} = a\Delta u - bu \tag{2}$$

is used in solid state physics, classical mechanics, nonlinear optics, and quantum field theory [6,7].

The time-fractional equation

$$\frac{\partial^\alpha u}{\partial t^\alpha} = a\Delta u - bu, \qquad 0 < \alpha \leq 2, \tag{3}$$

can be considered as the extension of the parabolic Equation (1) and hyperbolic Equation (2) and was studied in several publications [8–14].

It should be noted that such a generalization of many classical differential equations with integer derivatives has numerous applications in rheology, geology, physics, plasma physics, chemistry, geophysics, engineering, biology, bio-engineering, finance, and medicine (see [15–29], among many others). There is a great variety of inhomogeneous media where transport phenomena exhibit anomalous properties, the investigation of which is essential to refine understanding the basic characteristics of complex systems widely met in nature. Therefore, studying fractional equations has generated increasing attention of scientists in many disciplines. At present, the fractional diffusion-wave equation is generally used to describe a large class of systems at different scales (from the molecular [30] to the space one [31]) which cover media of the diverse nature (from plasma physics [29] to living tissue [3]). The study of this equation is also of interest from the point of view of understanding the complex spatio-temporal dynamics in nonlinear systems of fractional order [32,33].

In Equation (3) and further in this paper, for more concise notation, $\frac{d^\alpha f(t)}{dt^\alpha}$ denotes the Caputo fractional derivative [16,34]

$$\frac{d^\alpha f(t)}{dt^\alpha} = \frac{1}{\Gamma(n-\alpha)} \int_0^t (t-\tau)^{n-\alpha-1} \frac{d^n f(\tau)}{d\tau^n} \, d\tau, \qquad n-1 < \alpha < n, \tag{4}$$

and $\Gamma(\alpha)$ denotes the gamma function.

Ångström was the first to investigate the standard parabolic heat conduction equation under harmonic impact and laid the foundations for the new area of study known as "oscillatory diffusion" or "diffusion-waves" (see [35–37] and references therein). Periodic solutions of the bioheat equation were investigated in [38]. The harmonic point source in the bioheat equation was used in therapeutic hypotermia [39,40]; applications of the time-harmonic impact in ultrasound surgery were studied in [41].

As a rule, in the previous studies of diffusion or heat conduction equation the quasi-steady-state oscillations were investigated when the solution $u(\mathbf{x}, t)$ was represented as a product of a function of the spatial coordinates $U(\mathbf{x})$ and the time-harmonic term $e^{i\omega t}$ with the angular frequency ω

$$u(\mathbf{x}, t) = U(\mathbf{x}) \, e^{i\omega t} \tag{5}$$

without consideration of the initial conditions.

The use of assumption (5) is based on the well known formula for the derivative of the integer order n of the exponential function

$$\frac{d^n e^{\lambda t}}{dt^n} = \lambda^n \, e^{\lambda t}. \tag{6}$$

In the event of the non-integer order of time derivative, the assumption (5) cannot be used since [42]

$$\frac{d^\alpha e^{\lambda t}}{dt^\alpha} = \lambda^\alpha \, e^{\lambda t} \frac{\gamma(n-\alpha), \lambda t)}{\Gamma(n-\alpha)} \neq \lambda^\alpha \, e^{\lambda t}, \qquad n-1 < \alpha < n, \tag{7}$$

with $\gamma(a, x)$ being the incomplete gamma function [43]

$$\gamma(a, x) = \int_0^x e^{-u} u^{a-1} \, du. \tag{8}$$

It is worthy of notice that for the Riemann-Liouville fractional derivative [16,34] with the lower limit of integration at 0

$$D_{RL}^\alpha f(t) = \frac{d^n}{dt^n} \left[\frac{1}{\Gamma(n-\alpha)} \int_0^t (t-\tau)^{n-\alpha-1} f(\tau) \, d\tau \right], \qquad n-1 < \alpha < n, \tag{9}$$

we also have [16]

$$D_{RL}^\alpha e^{\lambda t} = t^{-\alpha} \, E_{1,1-\alpha} \, (\lambda t) \neq \lambda^\alpha \, e^{\lambda t}. \tag{10}$$

Here $E_{\alpha,\beta}(z)$ is the Mittag-Leffler function in two parameters α and β [16,34]

$$E_{\alpha,\beta} \, (z) = \sum_{n=0}^{\infty} \frac{z^n}{\Gamma(\alpha n + \beta)}, \qquad \Re(\alpha) > 0, \quad \beta \in C, \quad z \in C. \tag{11}$$

In this paper, the initial-boundary-value problem for Equation (3) is studied in a spherical domain for the case of central symmetry under the Dirichlet boundary condition varying harmonically in time. The present paper develops and extends the results of the previous investigations [44,45], where the corresponding problems for line and half-line domains were investigated.

2. Statement of the Problem

The time-fractional diffusion equation with mass absorption (mass release) is examined in a sphere

$$\frac{\partial^\alpha u}{\partial t^\alpha} = a \left(\frac{\partial^2 u}{\partial r^2} + \frac{2}{r} \frac{\partial u}{\partial r} \right) - bu, \quad 0 < r < R, \quad 0 < t < \infty, \quad 0 < \alpha \leq 2, \tag{12}$$

under zero initial conditions

$$t = 0: \quad u = 0, \quad 0 < \alpha \leq 2, \tag{13}$$

$$t = 0: \quad \frac{\partial u}{\partial t} = 0, \quad 1 < \alpha \leq 2, \tag{14}$$

and harmonic impact on the surface of a sphere

$$r = R: \quad u = u_0 e^{i\omega t}. \tag{15}$$

As in the case of classical diffusion equation (when $\alpha = 1$) and the wave equation (when $\alpha = 2$) the boundedness condition at the origin is also adopted:

$$r = 0: \quad u \neq \infty. \tag{16}$$

In what follows, the integral transform technique will be used. Recall the Laplace transform rule for the Caputo derivative

$$\mathcal{L} \left\{ \frac{d^\alpha f(t)}{dt^\alpha} \right\} = s^\alpha f^*(s) - \sum_{k=0}^{n-1} f^{(k)}(0^+) s^{\alpha - 1 - k}, \quad n - 1 < \alpha < n, \tag{17}$$

where the transform is marked by the asterisk, and s is the Laplace transform variable.

The following finite sin-Fourier transform is amenable to the central symmetric problem in a spherical domain $0 \leq r \leq R$ [46]. For the Dirichlet boundary condition:

$$\mathcal{F}\{f(r)\} = \tilde{f}(\xi_k) = \int_0^R r f(r) \frac{\sin(r\xi_k)}{\xi_k} \, dr, \tag{18}$$

$$\mathcal{F}^{-1}\{\tilde{f}(\xi_k)\} = f(r) = \frac{2}{R} \sum_{k=1}^{\infty} \xi_k \tilde{f}(\xi_k) \frac{\sin(r\xi_k)}{r}, \tag{19}$$

where the transform is marked by the tilde, and

$$\xi_k = \frac{k\pi}{R} \tag{20}$$

is the Fourier transform variable.

For the central symmetric Laplace operator

$$\mathcal{F} \left\{ \frac{d^2 f(r)}{dr^2} + \frac{2}{r} \frac{df(r)}{dr} \right\} = -\xi_k^2 \tilde{f}(\xi_k) + (-1)^{k+1} R f(R). \tag{21}$$

Applying to the problems (12)–(16) the Laplace transform with respect to time t and the finite sin-Fourier transform (18) with respect to the radial coordinate r, we get in the transform domain

$$\tilde{u}^*(\xi_k, s) = (-1)^{k+1} a R u_0 \frac{1}{s^\alpha + a\xi_k^2 + b} \frac{1}{s - i\omega}. \tag{22}$$

The solution is obtained after inversion of the integral transforms:

$$u(r,t) = \frac{2au_0}{r} \sum_{k=1}^{\infty} (-1)^{k+1} \xi_k \sin(r\xi_k) \int_0^t \tau^{\alpha-1} E_{\alpha,\alpha} \left[-\left(a\xi_k^2 + b \right) \tau^\alpha \right] e^{i\omega(t-\tau)} \, d\tau, \tag{23}$$

where $E_{\alpha,\beta}(z)$ is the Mittag-Leffler function (11), and the convolution theorem as well as the following equation for the inverse Laplace transform [16]

$$\mathcal{L}^{-1} \left\{ \frac{s^{\alpha-\beta}}{s^\alpha + \gamma} \right\} = t^{\beta-1} E_{\alpha,\beta} \left(-\gamma t^\alpha \right), \quad \alpha > 0, \quad \beta > 0, \tag{24}$$

have been used.

In numerical calculations, the nondimensional quantities are used:

$$\bar{u} = \frac{u}{u_0}, \qquad \bar{r} = \frac{r}{R}, \qquad \bar{t} = \frac{a^{1/\alpha}}{R^{2/\alpha}} t, \qquad \bar{\omega} = \frac{R^{2/\alpha}}{a^{1/\alpha}} \omega,$$

$$\bar{b} = \frac{R^2}{a} b, \qquad \eta_k = R\xi_k = k\pi. \tag{25}$$

Hence, using in integral in (23) the substitution $\tau = tw$, for the real part of the solution we get

$$\bar{u}(\bar{r}, \bar{t}, \bar{b}, \bar{\omega}) = \frac{2\bar{t}^\alpha}{\bar{r}} \sum_{k=1}^{\infty} (-1)^{k+1} \eta_k \sin(\bar{r}\eta_k) \int_0^1 w^{\alpha-1} E_{\alpha,\alpha} \left[-\left(\eta_k^2 + \bar{b} \right) \bar{t}^\alpha w^\alpha \right] \cos[\bar{\omega}\bar{t}(1-w)] \, dw, \tag{26}$$

where we can see that the solution $\bar{u}$ depends not only on time and spatial coordinate, but also on the parameters $\bar{b}$ and $\bar{\omega}$.

To simplify calculations, it would be worthwhile to introduce a substitution $z = w^\alpha$. Hence,

$$\bar{u}(\bar{r}, \bar{t}, \bar{b}, \bar{\omega}) = \frac{2\bar{t}^\alpha}{\bar{r}} \sum_{k=1}^{\infty} (-1)^{k+1} \eta_k \sin(\bar{r}\eta_k) \int_0^1 E_{\alpha,\alpha} \left[-\left(\eta_k^2 + \bar{b} \right) \bar{t}^\alpha z \right] \cos\left[\bar{\omega}\bar{t} \left(1 - z^{1/\alpha} \right) \right] \, dz. \tag{27}$$

To evaluate the Mittag-Leffler function the algorithms suggested in [47] were used; see also the MATLAB function [48] that implements these algorithms.

The numerical results are shown in Figures 1–5. Calculations were carried out for the grid size step $\Delta\bar{t} = 0.1$, $\Delta\bar{r} = 0.05$. A graphical representation of the solution (27) makes it possible to analyze not only the limiting cases of the problem (see Section 3), but also to understand the influence of the main parameters of the problem (including the value of the order of fractional derivative) on the spatial-temporal evolution of the solution.

It is seen from Figures that time oscillations of the solution are governed by the harmonic term $e^{i\omega t}$. The increasing of absorption parameter b decreases the oscillation amplitude (Figures 2b and 3d), whereas increasing α increases it (Figures 2a–d and 3d). As the Mittag-Leffler function

$$E_{2,2} \left(-x^2 \right) = \frac{\sin x}{x}, \tag{28}$$

the space oscillations of the solution depending on the order of fractional derivative appear for $\alpha \geq 1.5$ (Figures 1d and 2d) and become well-marked for α approaching 2. The influence of both factors is evident from Figures 4 and 5.

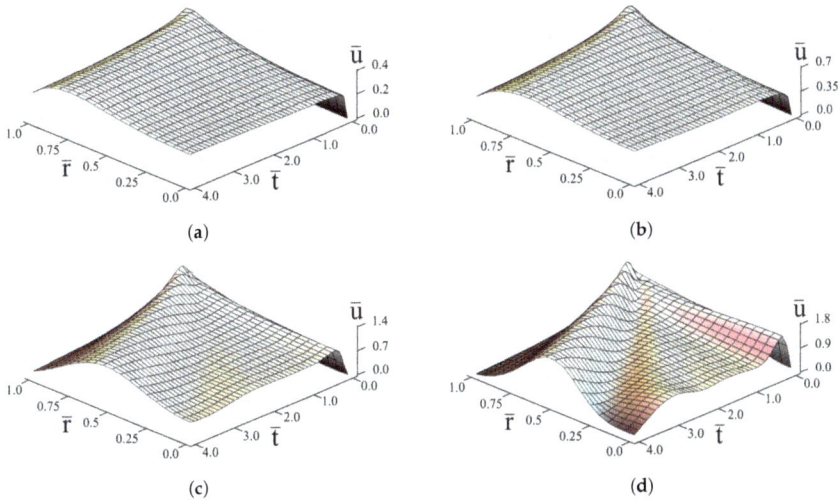

Figure 1. Evolution of the solution for the problems (12)–(15) under constant impact. The results of computer simulation of the formula (27) for the parameters $\bar{b} = 4$, $\bar{\omega} = 0$ and different values of α: $\alpha = 0.5$—(**a**); $\alpha = 0.75$—(**b**); $\alpha = 1.25$—(**c**); $\alpha = 1.5$—(**d**).

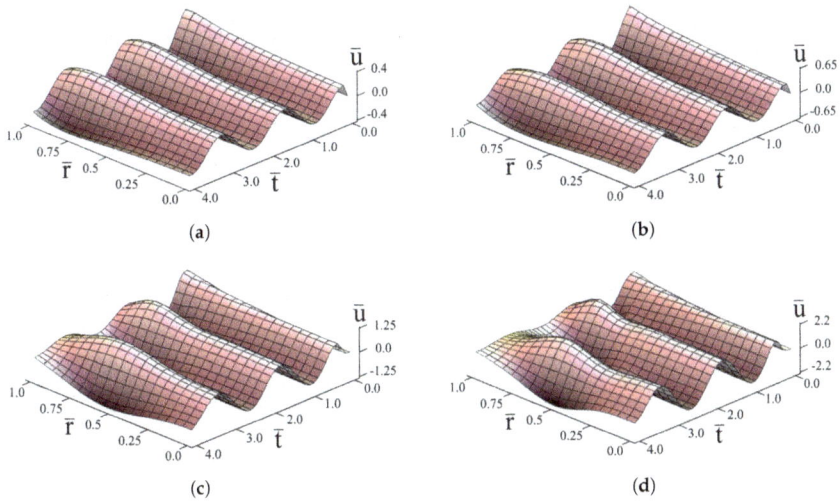

Figure 2. Evolution of the solution for the problems (12)–(15) under harmonic impact. The results of computer simulation of the formula (27) for the parameters $\bar{b} = 4$, $\bar{\omega} = 4$ and different values of α: $\alpha = 0.5$—(**a**); $\alpha = 0.75$—(**b**); $\alpha = 1.25$—(**c**); $\alpha = 1.5$—(**d**).

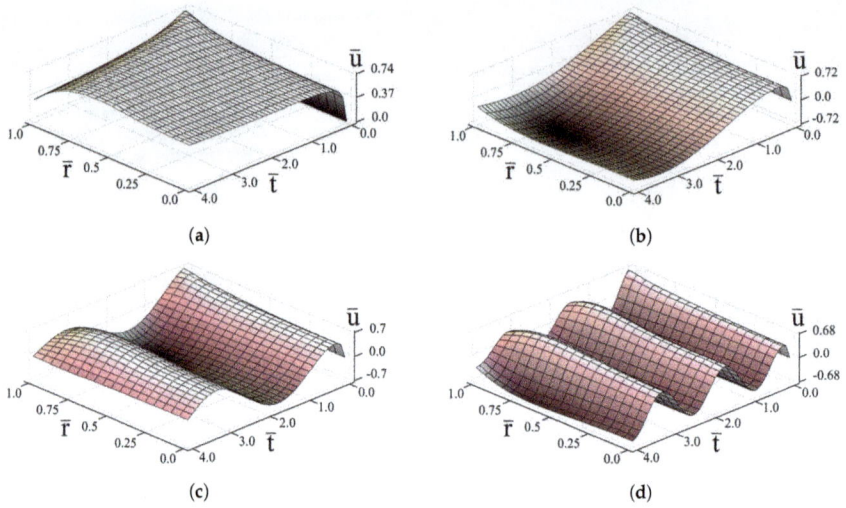

Figure 3. Evolution of the solution for the problems (12)–(15) with increasing the frequency $\bar{\omega}$ in the sub-diffusive case. The results of computer simulation of the formula (27) for the parameters $\bar{b} = 1$, $\alpha = 0.75$ and different values of $\bar{\omega}$: $\bar{\omega} = 0.0$—(**a**); $\bar{\omega} = 1.0$—(**b**); $\bar{\omega} = 2.0$—(**c**); $\bar{\omega} = 4.0$—(**d**).

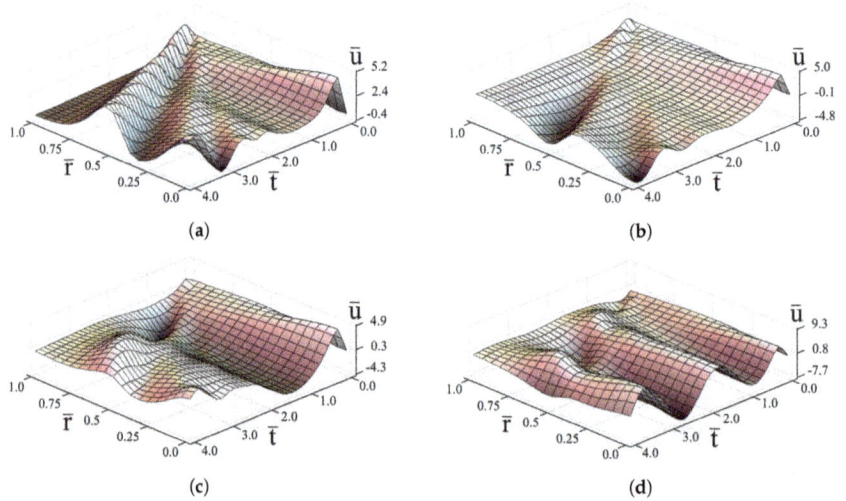

Figure 4. Evolution of the solution for the problems (12)–(15) with increasing the frequency $\bar{\omega}$ in the sub-wave case. The results of computer simulation of the formula (27) for the parameters $\bar{b} = 1$, $\alpha = 1.75$ and different values of $\bar{\omega}$: $\bar{\omega} = 0.0$—(**a**); $\bar{\omega} = 1.0$—(**b**); $\bar{\omega} = 2.0$—(**c**); $\bar{\omega} = 4.0$—(**d**).

(a)

(b)

(c)

(d)

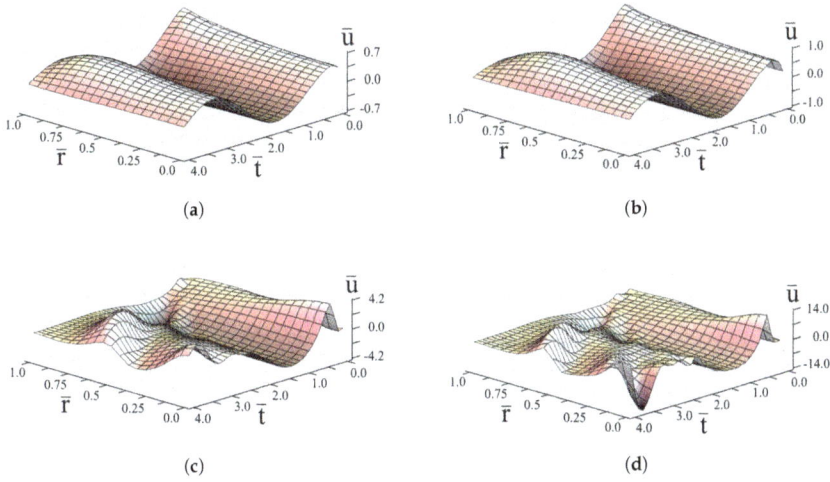

Figure 5. Evolution of the solution for the problems (12)–(15) for different orders of fractional derivative. The results of computer simulation of the formula (27) for the parameters $\bar{b} = 2$, $\bar{\omega} = 2.0$ and different values of α: $\alpha = 0.75$—(**a**); $\alpha = 0.95$—(**b**); $\alpha = 1.75$—(**c**); $\alpha = 1.95$—(**d**).

3. Analysis of the Quasi-Steady-State Oscillations

Now, we shall investigate two particular cases of the problem studied in the previous section corresponding to the integer values of the order of time derivative. For $\alpha = 1$, we have

$$\tilde{u}^*(\xi_k, s) = (-1)^{k+1} aRu_0 \frac{1}{s + a\xi_k^2 + b} \frac{1}{s - i\omega}. \tag{29}$$

Taking into account that [49]

$$\mathcal{L}^{-1}\left\{\frac{1}{(s+p)(s+q)}\right\} = \frac{e^{-qt} - e^{-pt}}{p - q}, \tag{30}$$

we arrive at the solution to the bioheat equation

$$u(r,t) = \frac{2au_0}{r} \sum_{k=1}^{\infty} (-1)^{k+1} \frac{\xi_k \sin(r\xi_k)}{a\xi_k^2 + b + i\omega} e^{i\omega t}$$

$$- \frac{2au_0}{r} \sum_{k=1}^{\infty} (-1)^{k+1} \frac{\xi_k \sin(r\xi_k)}{a\xi_k^2 + b + i\omega} e^{-(a\xi_k^2 + b)t}. \tag{31}$$

Similarly, for $\alpha = 2$,

$$\tilde{u}^*(\xi_k, s) = (-1)^{k+1} aRu_0 \frac{1}{s^2 + a\xi_k^2 + b} \frac{1}{s - i\omega}. \tag{32}$$

Taking into consideration that [49]

$$\mathcal{L}^{-1}\left\{\frac{1}{(s^2 + p^2)(s+q)}\right\} = \frac{1}{p^2 + q^2}\left[e^{-qt} - \cos(pt) + \frac{q}{p}\sin(pt)\right], \tag{33}$$

we obtain the solution to the Klein-Gordon equation

$$u(r,t) = \frac{2au_0}{r} \sum_{k=1}^{\infty} (-1)^{k+1} \frac{\xi_k \sin(r\xi_k)}{a\xi_k^2 + b - \omega^2} e^{i\omega t}$$

$$- \frac{2au_0}{r} \sum_{k=1}^{\infty} (-1)^{k+1} \frac{\xi_k \sin(r\xi_k)}{a\xi_k^2 + b - \omega^2} \left[\cos\left(\sqrt{a\xi_k^2 + b}\, t\right) + \frac{i\omega}{\sqrt{a\xi_k^2 + b}} \sin\left(\sqrt{a\xi_k^2 + b}\, t\right) \right]. \tag{34}$$

For integer α, we can assume that

$$u(r,t) = U(r)e^{i\omega t}. \tag{35}$$

For $\alpha = 1$, the function $U(r)$ fulfills the equation

$$\frac{d^2 U}{dr^2} + \frac{2}{r}\frac{dU}{dr} - \frac{b + i\omega}{a} U = 0, \tag{36}$$

under the boundary condition

$$r = R: \quad U(r) = u_0, \tag{37}$$

and for $b > 0$ has the solution bounded at the origin

$$U(r) = \frac{Ru_0}{r} \frac{\sinh\left[r\sqrt{(b + i\omega)/a}\right]}{\sinh\left[R\sqrt{(b + i\omega)/a}\right]}. \tag{38}$$

Therefore,

$$u(r,t) = \frac{Ru_0}{r} \frac{\sinh\left[r\sqrt{(b + i\omega)/a}\right]}{\sinh\left[R\sqrt{(b + i\omega)/a}\right]} e^{i\omega t} \tag{39}$$

(for negative value of b, sinh will be substituted by sin).

The first term in the solution (31) can be evaluated analytically using the following formula [50]

$$\sum_{k=1}^{\infty} (-1)^{k+1} \frac{k}{k^2 + p^2} \sin(kr) = \frac{\pi}{2} \frac{\sinh(rp)}{\sinh(\pi p)}, \quad -\pi < r < \pi,. \tag{40}$$

$$\sum_{k=1}^{\infty} (-1)^{k+1} \frac{k}{k^2 - p^2} \sin(kr) = \frac{\pi}{2} \frac{\sin(rp)}{\sin(\pi p)}, \quad -\pi < r < \pi. \tag{41}$$

It was emphasized in [51] that Equation (41) is also valid for complex values of p and hence turns into Equation (40) for imaginary p.

Taking into account Equation (40), we obtain that the first term in the solution (31) coincides with the quasi-steady-state solution (39), whereas the second term in Equation (31) describes the transient process.

The similar analysis can be carried out for $\alpha = 2$ based on the assumption (35). In this case, the function $U(r)$ fulfills the equation

$$\frac{d^2 U}{dr^2} + \frac{2}{r}\frac{dU}{dr} - \frac{b - \omega^2}{a} U = 0, \tag{42}$$

under the boundary condition

$$r = R: \quad U(r) = u_0, \tag{43}$$

and for $b > \omega^2$ has the solution bounded at the origin

$$U(r) = \frac{Ru_0}{r} \frac{\sinh\left[r\sqrt{(b-\omega^2)/a}\right]}{\sinh\left[R\sqrt{(b-\omega^2)/a}\right]}, \tag{44}$$

whereas for $b < \omega^2$

$$U(r) = \frac{Ru_0}{r} \frac{\sin\left[r\sqrt{(\omega^2-b)/a}\right]}{\sin\left[R\sqrt{(\omega^2-b)/a}\right]}. \tag{45}$$

Hence, for $b > \omega^2$

$$u(r,t) = \frac{Ru_0}{r} \frac{\sinh\left[r\sqrt{(b-\omega^2)/a}\right]}{\sinh\left[R\sqrt{(b-\omega^2)/a}\right]} e^{i\omega t}. \tag{46}$$

and for $b < \omega^2$

$$u(r,t) = \frac{Ru_0}{r} \frac{\sin\left[r\sqrt{(\omega^2-b)/a}\right]}{\sin\left[R\sqrt{(\omega^2-b)/a}\right]} e^{i\omega t}. \tag{47}$$

The first term in the solution (34) after accounting for Equations (40) and (41) coincides with the quasi-steady-state solutions (46) and (47), respectively, whereas the second term in (34) describes the transient process.

4. Conclusions

The time-fractional diffusion-wave equation with the Caputo fractional derivative of the order $0 < \alpha \leq 2$ with mass absorption was studied in a spherical domain under the Dirichlet boundary condition varying harmonically in time. The Caputo derivative of the exponential function has a much more complicated form than the corresponding derivative of the integer order. Hence, the assumption that the solution of the problem can be represented as a product of a function of the spatial coordinate and the time-harmonic term without consideration of the initial conditions cannot be used. The solution is obtained using the Laplace transform with respect to time and the finite sin-Fourier transform specifically adapted for a spherical domain and is expressed in terms of the Mittag-Leffler function. A graphical representation of the obtained analytical solution demonstrates the influence of the main parameters of the problem including the value of the order of fractional derivative on the spatial-temporal evolution of the solution.

Author Contributions: All authors have equally contributed to this work. All authors read and approved the final manuscript.

Funding: See acknowledgements of support below. The APC was funded by APVV-14-0892.

Acknowledgments: The first author is thankful for support from Rzeszow University of Technology (grant DS.FD.18.001). The second author acknowledges support provided by grants APVV-14-0892, APVV-18-0526, VEGA 1/0365/19, SK-SRB-18-0011, SK-AT-2017-0015, ARO W911NF-15-1-0228, COST CA15225. The third author would like to acknowledge the support of Jan Dlugosz University in Czestochowa (grant DS/WMP/6011/2018).

Conflicts of Interest: The authors declare no conflict of interest.

References

1. Crank, J. *The Mathematics of Diffusion*, 2nd ed.; Clarendon Press: Oxford, UK, 1975.
2. Pennes, H.H. Analysis of tissue and arterial blood temperatures in the resting human forearm. *J. Appl. Physiol.* **1948**, *1*, 93–122. [CrossRef] [PubMed]
3. Gafiychuk, V.V.; Lubashevsky I.A.; Datsko, B.Y. Fast heat propagation in living tissue caused by branching artery network. *Phys. Rev. E* **2005**, *72*, 051920. [CrossRef]

4.   Datsko, B.Y.; Gafiychuk, V.V.; Lubashevsky, I.A.; Priezzhev, A.V. Self-localization of laser-induced tumor coagulation limited by heat diffusion through active tissue. *J. Med. Eng. Technol.* **2006**, *30*, 390–396. [CrossRef]

5.   Polyanin, A.D. *Handbook of Linear Partial Differential Equations for Engineers and Scientists*; Chapman & Hall/CRC: Boca Raton, FL, USA, 2002.

6.   Gravel, P.; Gauthier, C. Classical applications of the Klein-Gordon equation. *Am. J. Phys.* **2011**, *79*, 447–453. [CrossRef]

7.   Wazwaz, A.-M. *Partial Differential Equations and Solitary Waves Theory*; Higher Education Press: Beijing, China; Springer: Berlin, Germany, 2009.

8.   Abuteen, A.; Freihat, A.; Al-Smadi, M.; Khalil, H.; Khan, R.A. Approximate series solution of nonlinear, fractional Klein-Gordon equations using fractional reduced differential transform method. *J. Math. Stat.* **2016**, *12*, 23–33. [CrossRef]

9.   Damor, R.S.; Kumar, S.; Shukla, A.K. Solution of fractional bioheat equation in terms of Fox's H-Function. *SpringerPlus* **2016**, *5*, 1–10. [CrossRef]

10.   Ferrás, L.L.; Ford, N.J.; Morgado, M.L.; Nóbrega, J.M.; Rebelo, M.S. Fractional Pennes' bioheat equation: theoretical and numerical studies. *Fract. Calc. Appl. Anal.* **2015**, *18*, 1080–1106. [CrossRef]

11.   Golmankhaneh, A.K.; Golmankhaneh, A.K.; Baleanu, D. On nolinear fractional Klein-Gordon equation. *Signal Process.* **2011**, *91*, 446–451. [CrossRef]

12.   Kheiri, H.; Shahi, S.; Mojaver, A. Analytical solutions for the fractional Klein-Gordon equation. *Comput. Meth. Diff. Equ.* **2014**, *2*, 99–114.

13.   Qin, Y.; Wu, K. Numerical solution of fractional bioheat equation by quadratic spline collocation method. *J. Nonlinear Sci. Appl.* **2016**, *9*, 5061–5072. [CrossRef]

14.   Vitali, S.; Castellani, G.; Mainardi, F. Time fractional cable equation and applications in neurophysiology. *Chaos Solitons Fractals* **2017**, *102*, 467–472. [CrossRef]

15.   Gorenflo, R.; Mainardi, F. Fractional calculus: Integral and differential equations of fractional order. In *Fractals and Fractional Calculus in Continuum Mechanics*; Springer: Wien, Austria, 1997; pp. 223–276.

16.   Podlubny, I. *Fractional Differential Equations*; Academic Press: San Diego, CA, USA, 1999.

17.   Povstenko, Y. Fractional heat conduction equation and associated thermal stresses. *J. Therm. Stress.* **2005**, *28*, 83–102. [CrossRef]

18.   Magin, R.L. *Fractional Calculus in Bioengineering*; Begell House Publishers, Inc.: Redding, CA, USA, 2006.

19.   Gafiychuk, V.V.; Datsko, B.Y. Spatiotemporal pattern formation in fractional reaction-diffusion systems with indices of different order. *Phys. Rev. E* **2008**, *77*, 066210. [CrossRef]

20.   Mainardi, F. *Fractional Calculus and Waves in Linear Viscoelasticity: An Introduction to Mathematical Models*; Imperial College Press: London, UK, 2010.

21.   Tarasov, V.E. *Fractional Dynamics: Applications of Fractional Calculus to Dynamics of Particles, Fields and Media*; Springer: Berlin/Heidelberg, Germany, 2010.

22.   Datsko, B.; Luchko, Y.; Gafiychuk, V. Pattern formation in fractional reaction-diffusion systems with multiple homogeneous states. *Int. J. Bifurcat. Chaos* **2012**, *22*, 1250087. [CrossRef]

23.   Datsko, B.; Gafiychuk, V. Complex nonlinear dynamics in subdiffusive activator-inhibitor systems. *Commun. Nonlinear Sci. Numer. Simul.* **2012**, *17*, 1673–1680. [CrossRef]

24.   Uchaikin, V.V. *Fractional Derivatives for Physicists and Engineers*; Springer: Berlin, Germany, 2013.

25.   Atanacković, T.M.; Pilipović, S.; Stanković, B.; Zorica, D. *Fractional Calculus with Applications in Mechanics: Vibrations and Diffusion Processes*; John Wiley & Sons: Hoboken, NJ, USA, 2014.

26.   Herrmann, R. *Fractional Calculus: An Introduction for Physicists*, 2nd ed.; World Scientific: Singapore, 2014.

27.   Povstenko, Y. *Fractional Thermoelasticity*; Springer: New York, NY, USA, 2015.

28.   Datsko, B.; Gafiychuk, V.; Podlubny, I. Solitary travelling auto-waves in fractional reaction–diffusion systems. *Commun. Nonlinear Sci. Numer. Simul.* **2015**, *23*, 378–387. [CrossRef]

29.   Anderson, J.; Moradi, S.; Rafiq, T. Non-linear Langevin and fractional Fokker-Planck equations for anomalous diffusion by Lévy stable processes. *Entropy* **2018**, *20*, 760. [CrossRef]

30.   Weiss, M.; Nilsson, T. In a mirror dimly: Tracing the movements of molecules in living cells. *Trends Cell Biol.* **2004**, *14*, 267–273. [CrossRef]

31.   Zelenyi, L.M.; Milovanov, A.V. Fractal topology and strange kinetics: From percolation theory to problems in cosmic electrodynamics. *Phys. Uspekhi* **2004**, *47*, 809–852. [CrossRef]

32. Gafiychuk, V.; Datsko, B. Different types of instabilities and complex dynamics in reaction-diffusion systems with fractional derivatives. *J. Comp. Nonlinear Dyn.* **2012**, *7*, 031001. [CrossRef]
33. Datsko, B.; Gafiychuk, V. Complex spatio-temporal solutions in fractional reaction-diffusion systems near a bifurcation point. *Fract. Calc. Appl. Anal.* **2018**, *21*, 237–253. [CrossRef]
34. Kilbas, A.A.; Srivastava, H.M.; Trujillo, J.J. *Theory and Applications of Fractional Differential Equations*; Elsevier: Amsterdam, The Netherlands, 2006.
35. Mandelis, A. Diffusion waves and their uses. *Phys. Today* **2000**, *53*, 29–33. [CrossRef]
36. Mandelis, A. *Diffusion-Wave Fields: Mathematical Methods and Green Functions*; Springer: New York, NY, USA, 2001.
37. Vrentas, J.S.; Vrentas, C.M. *Diffusion and Mass Transfer*; CRC Press: Boca Raton, FL, USA, 2013.
38. Lakhssassi, A.; Kengne, E.; Semmaoui, H. Modifed Pennes' equation modelling bio-heat transfer in living tissues: analytical and numerical analysis. *Natl. Sci.* **2010**, *2*, 1375–1385. [CrossRef]
39. Kengne, E.; Lakhssassi, A.; Vaillancourt, R. Temperature distributions for regional hypothermia based on nonlinear bioheat equation of Pennes type: Dermis and subcutaneous tissues. *Appl. Math.* **2012**, *3*, 217–224. [CrossRef]
40. Fasano, A.; Sequeira, A. *Hemomath. The Mathematics of Blood*; Springer: Cham, Switzerland, 2017.
41. Malinen, M.; Huttunen, T.; Kaipio, J.P. Thermal dose optimization method for ultrasound surgery. *Phys. Med. Biol.* **2003**, *48*, 745–762. [CrossRef]
42. Povstenko, Y. Fractional heat conduction in a space with a source varying harmonically in time and associated thermal stresses. *J. Therm. Stress.* **2016**, *39*, 1442–1450. [CrossRef]
43. Abramowitz, M.; Stegun, I.A. (Eds.) *Handbook of Mathematical Functions with Formulas, Graphs and Mathematical Tables*; Dover: New York, NY, USA, 1972.
44. Povstenko, Y.; Kyrylych, T. Time-fractional diffusion with mass absorption under harmonic impact. *Fract. Calc. Appl. Anal.* **2018**, *21*, 118–133. [CrossRef]
45. Povstenko, Y.; Kyrylych, T. Time-fractional diffusion with mass absorption in a half-line domain due to boundary value of concentration varying harmonically in time. *Entropy* **2018**, *19*, 346. [CrossRef]
46. Povstenko, Y. *Linear Fractional Diffusion-Wave Equation for Scientists and Engineers*; Birkhäuser: New York, NY, USA, 2015.
47. Gorenflo, R.; Loutchko, J.; Luchko, Y. Computation of the Mittag-Leffler function and its derivatives. *Fract. Calc. Appl. Anal.* **2002**, *5*, 491–518.
48. Podlubny, I. Mittag-Leffler Function; Calculates the Mittag-Leffler Function with Desired Accuracy, MATLAB Central File Exchange, File ID 8738. Available online: www.mathworks.com/matlabcentral/fileexchange/8738 (accessed on 17 April 2019).
49. Erdélyi, A.; Magnus, W.; Oberhettinger, F.; Tricomi, F. *Tables of Integral Transforms*; McGraw-Hill: New York, NY, USA, 1954; Volume 1.
50. Prudnikov, A.P.; Brychkov,Y.A.; Marichev, O.I. *Integrals and Series, Volume 1: Elementary Functions*; Gordon and Breach Science Publishers: Amsterdam, The Netherlands, 1986.
51. Magnus, W.; Oberhettinger, F. *Formeln und Sätze für die Speziellen Funkttionen der Mathematischen Physik*, 2nd ed.; Springer: Berlin, Germany, 1948.

Σ *mathematics*

MDPI

Article

Fractional Order Complexity Model of the Diffusion Signal Decay in MRI

Richard L. Magin [1,*], Hamid Karani [2], Shuhong Wang [3] and Yingjie Liang [3]

[1]   Department of Bioengineering at University of Illinois at Chicago, Chicago, IL 60607, USA
[2]   Department of Engineering Sciences and Applied Mathematics, Northwestern University,
     Evanston, IL 60208, USA; hamid.karani@northwestern.edu
[3]   Institute of Soft Matter Mechanics, College of Mechanics and Materials, Hohai University,
     Nanjing 211100, China; shuhong@hhu.edu.cn (S.W.); liangyj@hhu.edu.cn (Y.L.)
*   Correspondence: rmagin@uic.edu ; Tel.: +1-312-413-5528

Received: 11 March 2019; Accepted: 8 April 2019; Published: 12 April 2019

Abstract: Fractional calculus models are steadily being incorporated into descriptions of diffusion in complex, heterogeneous materials. Biological tissues, when viewed using diffusion-weighted, magnetic resonance imaging (MRI), hinder and restrict the diffusion of water at the molecular, sub-cellular, and cellular scales. Thus, tissue features can be encoded in the attenuation of the observed MRI signal through the fractional order of the time- and space-derivatives. Specifically, in solving the Bloch-Torrey equation, fractional order imaging biomarkers are identified that connect the continuous time random walk model of Brownian motion to the structure and composition of cells, cell membranes, proteins, and lipids. In this way, the decay of the induced magnetization is influenced by the micro- and meso-structure of tissues, such as the white and gray matter of the brain or the cortex and medulla of the kidney. Fractional calculus provides new functions (Mittag-Leffler and Kilbas-Saigo) that characterize tissue in a concise way. In this paper, we describe the exponential, stretched exponential, and fractional order models that have been proposed and applied in MRI, examine the connection between the model parameters and the underlying tissue structure, and explore the potential for using diffusion-weighted MRI to extract biomarkers associated with normal growth, aging, and the onset of disease.

Keywords: anomalous diffusion; complexity; magnetic resonance imaging; fractional calculus

---

1. Introduction

"One of the principal objects of theoretical research in my department of knowledge is to find the point of view from which the subject appears in the greatest simplicity". J. Willard Gibbs, Letter in Proc. Amer. Acad. Arts & Sci. (1881), pp. 420–421.

Mathematical models are a prism through which we can view the complexity of nature [1]. Just as a prism separates sunlight into the colors of its optical spectrum—portraying hidden features (frequency, intensity, and polarization)—the formulation of a model identifies features not displayed in the raw data. Parameter extraction and estimation mimic a spectrograph by selecting individual spectral components for analysis. In both measurement and modeling, we seek to isolate specific aspects of a physical phenomenon for further study. Success is measured by the degree to which the spectra (or model) captures particular features of the image or visual scene. In paint, the pigments can be identified by a UV-Vis spectrophotometer, but in painting color is described by hue, value, tone, tint, shade, and saturation, all of which are processed by the eye as characteristics of a picture. In modeling biomedical images, we use mathematics to help the brain recognize patterns by identifying new measures of image complexity that convey information in terms of contrast and resolution. Like a spectrograph attached to a telescope or microscope, a mathematical model seeks to capture information hidden in the image of a star or a cell.

Magnetic resonance imaging (MRI) systems are imaging spectrographs that combine camera and spectrometer so as to display the internal structure and composition of the human body [2]. Due to the mismatch between the sub-millimeter resolution of MRI and the sub-micron architecture of biological tissues, mathematical models are needed to describe the mesoscale complexity of living systems. Here, the versatility of MRI offers a variety of tools that encode the mobility of water (molecular rotation and translation) in terms of the magnetic resonance spectrum, and the decay or relaxation of its individual components [3]. In addition, manipulation of the imaging pulse sequence—through modulation of the applied radiofrequency and pulse gradient fields—provides image contrast. Interpretation of this sub-pixel (or for a selected slice, sub-voxel) contrast requires a dynamic model of the local magnetic dipole moment per unit volume, which in the case of diffusion, is the Bloch-Torrey equation [4].

Diffusion-weighted MRI (DW-MRI), based on the Bloch-Torrey equation, is implemented by selective phase encoding within each imaging voxel. This typically involves using a pair of rectangular gradient pulsed (Stejskal-Tanner pulses [5]) to capture the diffusion of water in tissue over the distance of several microns. The DW-MRI signal appears as an exponential signal decay (after appropriate normalization for the local signal intensity and the intrinsic T_1 and T_2 relaxation times) [6]. Hence, the connection between the molecular diffusion coefficient, D (mm^2/s) and the detected signal decay involves coarse graining of the magnetic dipole moment per unit volume (M, Amp/m), selection of the transverse component (M_{xy}), detection of an induced time domain signal ($S(t)$, volts/s), slice selection, phase and spatial encoding to form an image, $I(x,y)$, and finally diffusion encoding into the diffusion signal for each imaging voxel, $S(b) = S_0 \exp(-bD)$, where diffusion-weighting, b, depends on the pulse gradient strength, g, width, δ, and separation time, Δ, for rectangular Stejskal-Tanner pulses, e.g., $b = (\gamma g \delta)^2 (\Delta - \delta/3)$, and γ is the gyromagnetic ratio, 42.57 MHz/Tesla for water protons. The analytical solution of the Bloch-Torrey equation is only possible for simple geometries and relatively uniform samples (ideally for Gaussian diffusion in homogenous, isotropic, and unbounded materials) [7]. Relaxation of these conditions leads to the need for additional gradient pulses, which extend the overall imaging time. Hence, the need to simplify this model is guided, on one hand, by a desire to correlate the detected DW-MRI signal with specific tissue features (e.g., isotropic versus anisotropic diffusion or normal versus anomalous diffusion) and on the other by the need to keep the imaging protocol as short as possible—at least in clinical exams.

Anomalous diffusion is characterized by a non-linear growth in the mean squared displacement (MSD) with time [8]. In complex, heterogeneous materials, such as biological tissues, anomalous diffusion has been observed directly in cells and membranes using high resolution optical techniques (fluorescence correlation spectroscopy (FCS), and fluorescence recovery after photobleaching (FRET)), and indirectly in tissues using DW-MRI [9]. Molecular crowding and close cell packing hinder the movement of water reducing the apparent diffusion coefficient from its nominal value of 2.5×10^{-3} mm^2/s to 0.8×10^{-3} mm^2/s in brain gray matter, while the MSD remains proportional to the diffusion time [10]. At longer diffusion times (typically greater than 50 ms), the water trapped in and around cells finds its motion not only hindered, but also restricted [11]. For example, water trapped in cells and sub-cellular organelles crosses the associated membranes very slowly and this barrier alters the free diffusion and water transport such that the MSD now increases with time in a sub-linear manner, MSD = $2Dt^\alpha$, where $0 < \alpha < 1$; a process termed anomalous and sub-diffusive. Both Gaussian (fractional Brownian motion, Langevin equation) and non-Gaussian (continuous time random walk, (CTRW), Lorentz obstructed motion) diffusion models have been suggested describe the DW-MRI data [12]. An example of DW-MRI data is displayed in Figure 1, which is a plot of the normalized signal intensity (S/S_0) versus b-value for white matter (WM), gray matter (GM), and cerebral spinal fluid (CSF) acquired from a normal human brain using a clinical 3 Tesla MRI scanner [13]. The semilogarithmic plot for CSF is the expected straight line characteristic of a single exponential decay. The GM signal exhibits a smaller apparent diffusion coefficient (decay rate) with a small upward curvature at *b*-values above 1000 s/mm². Hence, both CSF and GM can be described as examples of normal Gaussian diffusion. The WM signal, however, shows a much larger curvature in this plot and a decay rate (slope) that appears to decrease with increasing *b*-value [14]. Water diffusion in WM is an example of anomalous diffusion where the highly myelinated axons and dense fiber bundles of the white matter add barriers that restrict the movement of tissue water.

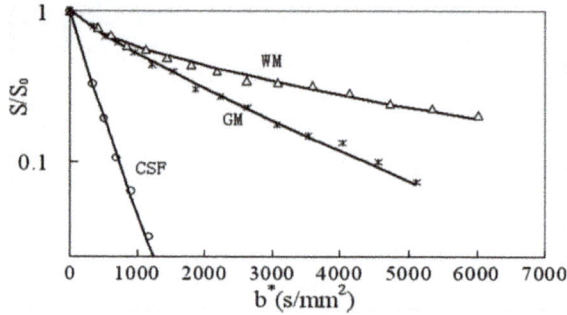

Figure 1. Normalized signal intensity plotted versus b for selected regions of interest (ROI) in white matter (WM), gray matter (GM) and cerebrospinal fluid (CSF) for a human brain (reprinted from ref. [13] with permission). DW-EPI T2-weighted image at 3.0 Tesla, TR/TE = 4000/97 ms, FOV = 22 cm · 22 cm, matrix 128 ·72 (zero padded to 256 · 256 during image reconstruction), in plane resolution = 1.72 mm × 3.05 mm and slice thickness = 4 mm. The experimental data were fit to the fractional order stretched exponential model, S(b) = S_0exp[(−bD)$^\alpha$)], (WM, α = 0.64, D = 0.41 × 10^{-3} mm²/s ; GM, α = 0.82, D = 0.66 × 10^{-3} mm²/s ; CSF, α = 0.95, D = 2.72 × 10^{-3} mm²/s).

Two research paradigms are commonly explored in DW-MRI. First, one can build simplified models of tissue structure by selecting two, three, or four tissue compartments or components, each with a specific, and to be determined, parameter set (e.g., intra- and extra-cellular diffusion and membrane permeability). Diffusion Tensor Imaging (DTI) applies this approach to each image voxel via gradient pulses applied in multiple (at least six) directions. Other approaches (CHARMED, NOODI, AxCaliper, etc.) [15], extend the data collection to multiple *b*-value shells sampled in hundreds of directions. Alternatively, one can select a heuristic model for *S(b)*, such as the kurtosis, S(b) = S_0exp(−bD)exp[(K/6)(bD)²] [16], or the stretched exponential model, S(b) = S_0 exp[−(bD)$^\alpha$] [17], and fit the available clinical data to the extended set of model parameters {S_0,D,K} or {S_0,D,α}, respectively. Such models, since they have an extra parameter, will provide an improved fit to *S(b)* data that extends over multiple *b*-values. The sword of Damocles in the first approach is the added imaging time, while for the second, it is the dichotomy between the precision of the parameter fit and the ambiguity in the connection between the tissue composition and fitting parameter. Another factor, most often overlooked for both models is the likelihood that the fitting process is degenerate, that is, more than one set of parameters can fit the data with the same least squares fit error [18].

In this paper, we will combine the two modeling approaches described above by using a varying diffusion coefficient *D(b)* to account for the complexity of the tissue as an inverse power law for higher values of diffusion weighting (*b*-values), and by generalizing the stretched exponential and kurtosis diffusion decay models by extending the governing diffusion decay to fractional order. The fractional order model naturally accounts for both multiple compartments and non-Gaussian diffusion. After a few preliminary definitions, we will outline an extended form for the fractional diffusion coefficient model, and the corresponding integer and fractional order diffusion decay equations. Examples of the expected functional behavior of the diffusion signal decay as a function of *b*-value will be plotted and fit to ex vivo bovine optic nerve tissue. In the Discussion we will describe how these models expand the available modeling tools for DW-MRI data and describe how they can be used to fit animal and clinical data. Overall, this approach provides a way to extend the heuristic stretched exponential approach toward more complete multiple-compartment models. This bridge uses fractional order derivatives and varying diffusion coefficients as connecting links.

2. Definitions and Properties

The signals analyzed in DW-MRI, $f(b) = S(b)/S_0$, are expressed in terms of b, the diffusion-weighting variable defined above for a pair of rectangular diffusion pulses [2–4]. Typically, $f(b)$ is a single valued function that is positive, real, and monotonically decreasing for b-values in the range from 0 to 5000 s/mm^2. For each pixel in an MR image, $f(b)$ is fit to data acquired by varying b in magnitude and gradient direction. The default function used in most DW-MRI is the single exponential, $f(b) = \exp(-bD)$. This model reflects the underlying assumptions of a linear, first order relaxation model appropriate for free or hindered Gaussian diffusion. In biological tissues, a sum of exponentials or the stretched exponential, $f(b) = \exp[(-bD)^\alpha]$, where $0 < \alpha < 1$, is often needed to describe the data, particularly at high b-values. When neither of these approaches are satisfactory, non-Gaussian behavior is considered, and in this paper we will describe it by using a fractional-order relaxation model. Our model involves the Caputo fractional derivative, $^C D^\alpha[f(b)]$, so below we have defined this derivative operator and its properties. In addition, we also define the Mittag-Leffler function because this function frequently appears in various forms when solving fractional-order differential equations. We will show that the Mittag-Leffler function is also a generalization of the exponential.

2.1. Caputo Fractional Derivative

The simplest description of the non-local, Caputo fractional derivative of order α ($0 < \alpha < 1$) is as an integral convolution of a power law decay, $b^{-\alpha}/\Gamma(1 - \alpha)$, with $f(b)$ [19], hence:

$$
{}_0^C D_b^\alpha f(b) = f(b) * \frac{b^{-\alpha}}{\Gamma(1-\alpha)} \ , \tag{1}
$$

where the gamma function, $\Gamma(z)$ is defined for all $z > 0$ by:

$$
\Gamma(z) = \int_0^\infty x^{z-1} e^{-x} dx \ . \tag{2}
$$

For integer values of n, $\Gamma(1 + n) = n\Gamma(n) = n!$, and in the limit as α approaches 1, the Caputo fractional derivative converges to the integer result, $^C D^\alpha f(b) = df(b)/db$. Whereas, for $f'(b) = df/db$, we have:

$$
{}_0^C D_b^\alpha f(b) = \begin{cases} \frac{1}{\Gamma(1-\alpha)} \int_0^b \frac{1}{(b-b')^\alpha} \left(\frac{df(b')}{db'} \right) db' & 0 < \alpha < 1 \\ \frac{df(b)}{db} & \alpha = 1 \end{cases} \ , \tag{3}
$$

Applying the Caputo fractional derivative to monomial of arbitrary degree, κ yields:

$$
{}_0^C D_b^\alpha b^\kappa = \frac{\Gamma(1+\kappa)}{\Gamma(1+\kappa-\alpha)} (b^{\kappa-\alpha}) , \tag{4}
$$

which gives 1 and 0 for $b = 1$, and $\alpha = 0$ and $\alpha = 1$, respectively. Hence, the Caputo fractional derivative is a simple extension of the integer result $db^\kappa/db = [\kappa!/(\kappa - 1)!]b^{(\kappa-1)} = \kappa\, b^{(\kappa-1)}$. However, for other common functions, such as the $\sin(b)$, $\cos(b)$ and $\exp(b)$, the results are not quite as simple.

Applying the Caputo fractional derivative term by term to the power series expansion of the single exponential $f(b) = \exp(b)$, yields:

$$
{}_0^C D_b^\alpha \left(e^b \right) = b^\alpha \left(e^b \right) \left(1 - \frac{\Gamma(-\alpha b)}{\Gamma(-\alpha)} \right) . \tag{5}
$$

Rather than tabulate all the changes incurred when applying the Caputo fractional derivative to familiar functions, mathematicians have developed particular special functions with simpler behavior under the fractional derivative operation. One such function was defined by the Swedish mathematician, Magnus Gustaf Mittag-Leffer; it is a generalization of the exponential.

2.2. Mittag-Leffer Function

The Mittag-Leffer function, like the Bessel function, comes in different kinds (first, second, modified, etc.), and occurs in both integer and fractional order. It was developed out of the need to simplify the solution of certain classes of ordinary differential equations. Also, like the Bessel function, the Mittag-Leffer function can be defined in terms of a convergent power series. For more information on the Mittag-Leffer function, please consult the monograph by Mainardi and colleagues [20]. There are many versions of the Mittag-Leffler function which subsume almost all of the basic functions of mathematical physics. In this article, we only need three kinds of Mittag-Leffler functions, which we will define in terms of their corresponding power series.

2.3. Exponential Function

Since $\Gamma(n + 1) = n!$, the simple decaying exponential has the following power series representation:

$$\exp(-b) = \sum_{n=0}^{\infty} \frac{(-b)^n}{\Gamma(n+1)} \ . \tag{6}$$

2.4. One Parameter Mittag-Leffer Function

The one parameter Mittag-Leffler function [20] is a generalization of the exponential with the parameter, α, inserted a multiplying the integer n. It has the following power series representation:

$$E_\alpha(-b) = \sum_{n=0}^{\infty} \frac{(-b)^n}{\Gamma(n\alpha+1)} \ . \tag{7}$$

Note, that $E_1(-b) = \exp(-b)$ and $E_2(-b^2) = \cos(b)$.

2.5. Two Parameter Mittag-Leffer Function

The two parameter Mittag-Leffler function [20] includes a second parameter β in the gamma function of the power series as:

$$E_{\alpha,\beta}(-b) = \sum_{n=0}^{\infty} \frac{(-b)^n}{\Gamma(n\alpha+\beta)} \ . \tag{8}$$

Note that here, $E_{1,1}(-b) = \exp(-b)$, $E_{1,2}(-b) = [1 - \exp(-b)]/b$, and $E_{2,2}(-b^2) = [\sin(b)]/b$.

2.6. Three Parameter Mittag-Leffer Function

There are several ways to extend the Mittag-Leffler function to three parameters—one credited to Prabhakar [21], and another to Kilbas and Saigo [22]. We will use the Kilbas-Saigo definition, which in power law form can be written as:

$$E_{\alpha,m,\ l}(-b) = \sum_{n=0}^{\infty} C_n(-b)^n \ , \tag{9}$$

where $c_0 = 1$ and:

$$c_n = \prod_{j}^{n-1} \frac{\Gamma[\alpha(jm+l)+1]}{\Gamma[\alpha(jm+l+1)+1]} \ . \tag{10}$$

Note that $E_{1,1,1}(-b) = exp(-b)$, $E_{\alpha,1,1}(-b) = E_\alpha(-b)$, and $E_{\alpha,\beta,1}(-b) = E_{\alpha,\beta}(-b)$.

Two useful results can be established by applying the Caputo derivative term by term to these power series expressions. First, we find the following for the one parameter Mittag-Leffler function:

$$ {}_0^C D_b^\alpha E_\alpha\left[-(bD)^\alpha\right] = -D^\alpha E_\alpha\left[-(bD)^\alpha\right] . \tag{11} $$

Thus, the stretched Mittag-Leffler function solves the homogeneous fractional order differential equation ${}_0^C D_b^\alpha y(b) + D^\alpha y(b) = 0$. And, for $\alpha = 1$, the above expression reduces to the simple relationship $d[exp(-bD)]/db = -D\ exp(-bD)$, and gives the solution to the corresponding first order differential equation.

For the inhomogeneous fractional order differential equation:

$$ {}_0^C D_b^\alpha y(b) + D^\alpha y(b) = g(b), \tag{12} $$

with $0 < \alpha < 1$, and $y(0) = B$, the solution [22] is:

$$ y(b) = BE_\alpha\left[-(bD)^\alpha\right] + \int_0^b (b - b')^{\alpha-1} E_{\alpha,\alpha}\left[-(b'D)^\alpha g(b')db' \right. . \tag{13} $$

Finally, for the following homogeneous fractional order differential equation:

$$ {}_0^C D_b^\alpha y(b) + D^\alpha \beta (bD)^\beta y(b) = 0 , \tag{14} $$

and $y(0) = B$, the solution [23] is:

$$ y(b) = BE_{\alpha,1+\beta/\alpha,\beta/\alpha}\left[-(bD)^{\alpha+\beta}\right] , \tag{15} $$

which is a reduced form of the three parameter Mittag-Leffler function, with $\alpha = \alpha$, $m = 1 + \beta/\alpha$, and $l = \beta/\alpha$. Also, α and β are restricted to the conditions, $0 < \alpha < 1$, and $-\alpha < \beta < 1 - \alpha$. We will refer to this particular case as the Kilbas-Saigo function in the following text.

3. Results

3.1. Fractional Order D(b) Models

A key feature of the anomalous diffusion exhibited in DW-MRI is the fall-off or reduction of the decay rate at b-values above 1000 s/mm². A recent paper [14] surveyed integer order models of this phenomena, where the apparent diffusion coefficient D_0 was expressed as a decaying function of b with a separate decay rate, D_1, for example, $D(b) = D_0 exp[-(bD_1)]$. Here, we focus on fractional order models for systems with both a constant and an inverse power law decay rate. A summary of our approach is given in Figure 2, which illustrates the different integer and fractional order models that will be described. For the integer order case, with a constant diffusion coefficient, $D(b) = D_0$, the linear differential equation model, yields the classical exponential decay, $S(b)/S_0 = exp[-(bD_0)]$. A common extension of this model [23,24] assumes an inverse power law decay rate, $D(b) = (1 + \beta)D_0(bD_0)^\beta$, where $-1 < \beta < 0$. This integer order, linear differential equation has a stretched exponential solution, $S(b)/S_0 = exp[-(bD_0)^{(1+\beta)}]$, which has often been used to fit DW-MRI signals (see, [17,24,25]). The fractional β parameter is zero for the cases of normal and hindered diffusion, as in CSF and gray matter, but less than one for situations where anomalous diffusion appears—such as in brain white matter. Nevertheless, this signal decay model is still exponential, and will decay faster than a power law at high b-values.

Fractional-order

$$\frac{d^\alpha S(b)}{db^\alpha} + D_0^\alpha S(b) = 0$$

$$\frac{d^\alpha S(b)}{db^\alpha} + D_0^\alpha (bD_0)^\beta S(b) = 0$$

Model Selection for the Decay of Diffusion MRI Signal

$$\frac{dS(b)}{db} + D_0 S(b) = 0$$

Fractional-order with power law rate

Constant Diffusion Coefficient

$$\frac{dS(b)}{db} + (1+\beta)D_0\,(bD_0)^\beta S(b) = 0$$

Power law rate

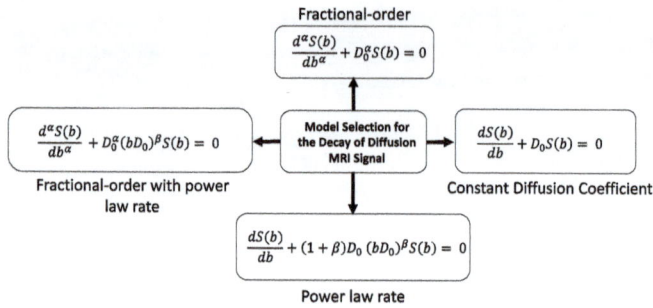

Figure 2. A diagram illustrating four approaches selected for generalization of the classical Gaussian model of diffusion-weighted signal attenuation in magnetic resonance imaging (MRI) (integer or fractional order, each with a constant diffusion coefficient or one with an inverse power law decay). An example of a representative differential equation for $S(b)$ is shown for each case.

Fractional order models asymptotically decay as a power law, so another way to capture the power law tail of the anomalous diffusion signal decay is to consider the decay process to be governed by a Caputo fractional order derivative (order α, with $0 < \alpha < 1$) and a constant diffusion coefficient, D_0. The solution for this case is the single parameter stretched Mittag-Leffer function, $S(b)/S_0 = E_\alpha[-(bD_0)^\alpha]$. This model is attractive because at low b-values it decays as a stretched exponential (approx. $1 - b^\alpha/\Gamma(1 + \alpha) + \dots$) and at high b-values decays as a power law of the form $(b^{-\alpha}/\Gamma(1 - \alpha))$ [23]. It has been fit to DW-MRI data by Magin, Ingo, and others [12,26]. For completeness, we also show the case of a fractional order Caputo derivative model (order α) with an inverse power law decay rate (order β). The solution for this case is a three parameter Mittag-Leffler function (Kilbas-Saigo form, see Equation (15) and Table 1), but for the inverse power law decay rate is expressed using only α and β as $S(b)/S_0 = E_{\alpha,1+\beta/\alpha,\beta/\alpha}[-(bD_0)^{\alpha+\beta}]$, [23]. Two things should be noted about this function. First, the conflation of the separate fractional parameters—the α connected with the order of the Caputo fractional derivative, while the β is connected with the presumed inverse power law decay of the diffusion decay rate, $D(b)$. The second thing to note is that, the α and β in this solution are not the α, β of the second order Mittag-Leffler function, but reflect a particular condition on the α, m, and l parameters of the three parameter Mittag-Leffler function. In addition, from its construction, we note that the fractional order derivative with the inverse power law decay rate case includes all of the other models as special cases. All four are displayed in Table 1 for reference and ease of comparison.

Table 1. Summary of Diffusion Decay functions, $S(b)$, for Selected Cases.

Constant	Power Law
$S(b) = S_0 \exp(-bD_0)$	$S(b) = S_0 \exp\left(-(bD_0)^{1+\beta}\right)$
Mittag-Leffler	Kilbas-Saigo
$S(b) = S_0 E_\alpha\left(-(bD_0)^\alpha\right)$	$S(b) = S_0 E_{\alpha,1+\frac{\beta}{\alpha},\frac{\beta}{\alpha}}\left(-(bD_0)^{\alpha+\beta}\right)$

$b = (\gamma\delta)^2\left(\Delta - \frac{\delta}{3}\right)$, $S(b = 0^+) = S_0$, where, α, β, and D_0 are constants, and $0 < \alpha < 1$, $-1 < \beta < 0$.

In order to display the behavior of the four signal decay models, we have prepared plots of the different diffusion signal decay functions for a selected set of the α and β parameters. The functions were evaluated using MATLAB® (MathWorks, Natick, MA 01760) and the code is publicly available at [27]. In the MRI field, the stretched exponential was first applied by Bennett et al. [17] to model DW-MRI data. Figure 3 is a plot $S(b)/S_0 = exp[-(bD_0)^{(1+\beta)}]$ versus b for b-values between 0 and 4000 s/mm^2, and $D_0 = 1 \times 10^{-3}$ mm^2/s. We selected five values for $(1 + \beta)$ = 0.2, 0.4, 0.6, 0.8, and 1.0. The last value in this sequence, corresponding to $\beta = 0$, gives the single exponential function. Three things should be noted about these normalized decay curves. First, they all go through the point $f(b) =$

0.37 when $bD_0 = 1$, and for smaller b-values, all the curves fall below the exponential, while for $bD_0 > 1$, the decay curves are all higher. Thus, the stretched exponential function mimics the combination of a fast plus a slow decay. The deviation from the exponential decreases as the value of $(1 + \beta)$ increases. Second, slope of the stretched exponential becomes increasingly steep near $b = 0$, and in fact diverges when $(1 + \beta) < 1$. This behavior can be eliminated by shifting the stretched exponential, but that changes the character of the decay [28]. Third, for high b-values, the stretched exponential decays as a power law on a semilogarithmic graph, but not on a linear scale. The multiexponential character of the stretched exponential can be extracted using the inverse Laplace transform (see the work of Berberan-Santos et al., [29]) and it also finds applications in the cumulative function of the Weibull probability distribution [30].

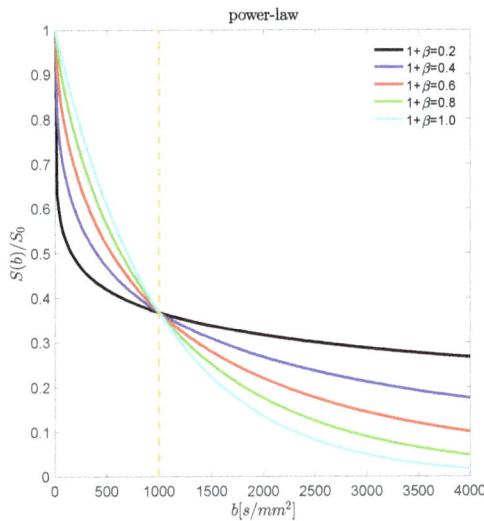

Figure 3. The normalized signal decay of $S(b)/S_0$ for a stretched exponential decay is plotted for b-values between 0 and 4000 s/mm^2, and $D_0 = 1 \times 10^{-3}$ mm^2/s. Five values for $(1 + \beta) = 0.2, 0.4, 0.6, 0.8$, and 1.0 are displayed. The last value in this sequence gives the single exponential function. As the value of $(1 + \beta)$ moves toward 1, the shape of the decay curve transitions from a multiple exponential with fast and slow components to a single exponential decay.

The single parameter Mittag-Leffler function is well known as a relaxation model for dielectric and viscoelastic materials [31,32], and more recently in MRI [12,26]. is a plot $S(b)/S_0 = E_\alpha[-(bD_0)^\alpha]$ versus b for b-values between 0 and 4,000 s/mm^2, and $D_0 = 1 \times 10^{-3}$ mm^2/s. We selected five values for $\alpha = 0.2, 0.4, 0.6, 0.8$, and 1.0. The last value in this sequence, corresponding to $\alpha = 1$, gives the single exponential function. The one parameter Mittag-Leffler function behaves, overall, in a manner similar to that of the stretched exponential—falling below the exponential at low b-values, and above it at high b-values. However, as shown in Figure 4, all the curves do not pass through $bD_0 = 1$, but most go through $S(b)/S_0 = 0.5$ near $bD_0 = 0.7$. And, for high b-values, on linear scales, the curves flatten out as the α values decrease, reflecting a significant reduction in the apparent diffusion constant—a hallmark of restricted diffusion. On logarithmic scales at high b-values, the one parameter Mittag-Leffler will appear as a straight line, which is characteristic of a pure power decay (see, Carpinteri and Mainardi [31] for these plots). The one parameter Mittag-Leffler has been used to fit DW-MRI data by several groups [12,26] and there is a growing recognition that the fractional order, α, is a homogeneity measure of the sub-voxel tissue environment—as the cellular heterogeneity increases (homogeneity decreases), the α values fall below 1.0. This view is consistent with the spectral distribution model of the one parameter Mittag-Leffler presented by Carpinteri and Mainardi [31], and the inverse Laplace transform analysis of Berberan-Santos [33]. Both models coalesce to a Dirac delta function when $\alpha = 1.0$, and broaden significantly as $\alpha < 1.0$, indicating a growing population of compartments with low diffusion rates (coefficients). Hence, one can interpret the one parameter Mittag-Leffler function as a concise multiexponential model of complex, heterogeneous material, and a natural candidate to use when searching for imaging biomarkers of tissue changes or pathology.

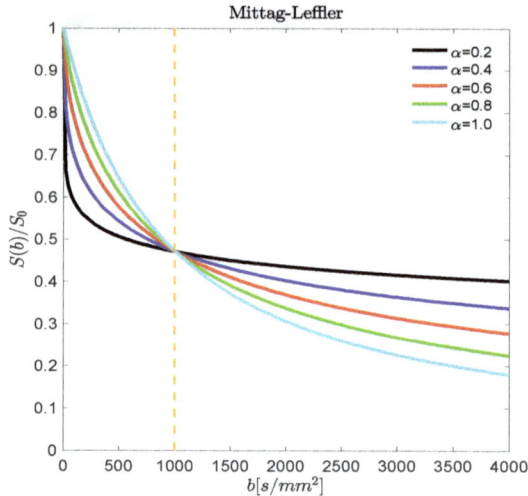

Figure 4. The normalized stretched Mittag-Leffler signal decay of $S(b)/S_0$ is plotted versus b-values between 0 and 4000 s/mm^2, with $D_0 = 1 \times 10^{-3}$ mm^2/s. Five values for $\alpha = 0.2, 0.4, 0.6, 0.8$, and 1.0 are displayed. The last value in this sequence, corresponding to $\alpha = 1$, gives the single exponential function. As the value of α decreases, the shape of the decay curve transitions from stretched exponential to an asymptotic power law decay.

The three parameter Mittag-Leffler function (Kilbas-Saigo function), when expressed using only α and β was derived and first applied in MRI by Hanyga and Seredynska [34], but plotted only over a narrow range. Here, we follow the recent work of Capelas de Oliveira, Mainardi, and Vaz [22] and, in addition, suggest that the α and β parameters of the Kilbas-Saigo function might provide separate tissue characterization parameters. We selected different values for α and β subject to the conditions, $0 < \alpha < 1$ and $0 < \alpha + \beta < 1$ to ensure complete monotonicity and non-negativity of the spectral distribution [22]. Figure 5 is a plot of $S(b)/S_0 = E_{\alpha,1+\beta/\alpha,\beta/\alpha}[-(bD_0)^{\alpha+\beta}]$ versus b for b-values between 0 and 4000 s/mm^2, and $D_0 = 1 \times 10^{-3}$ mm^2/s with $\alpha = 1$, and $\beta = 0, -0.2, -0.4, -0.6$, and -0.8. The signal decay in this case is a weighted stretched exponential and is similar to the stretched exponential case shown in Figure 3 (i.e., $S(b)/S_0 = exp[-(1+\beta)^{-1}(bD_0)^{1+\beta}]$). Figure 6 is a plot of the Kilbas-Saigo function for the same range of b-values and D_0, but with $\alpha = 0.8$, and $\beta = 0.2, 0, -0.2, -0.4$, and -0.6. In this figure we observe the effects of both α and β on the decay. The fractional derivative order α contributes multiexponential features to the decay, while the power law exponent β influences the low b-values more strongly than the high b-values. Figure 7 is a plot of the Kilbas-Saigo function for the same range of b-values and D_0, but with $\beta = 0.2$, and $\alpha = 0.2, 0.4, 0.6, 0.8$, and 1.0. In the figure, we see for a fixed power law decay of $D(b)$ the shift from lower to higher decay rates as α approaches the exponential function. The separate effects of the two fractional order parameters are also portrayed in the spectral distribution (inverse Laplace transform) of the Kilbas-Saigo function, where for $\beta = 0$ (Figure 4 in [22]) decreasing α shifts the spectrum to lower frequencies (smaller D values), which is the basis for the power law tail in the signal decay curve. On the other hand, for fixed α (Figures 5 and 6 in [22]), decreasing β only broadens the distribution about its mean value, which attenuates the diffusion signal in a more uniform manner over the entire range of b-values.

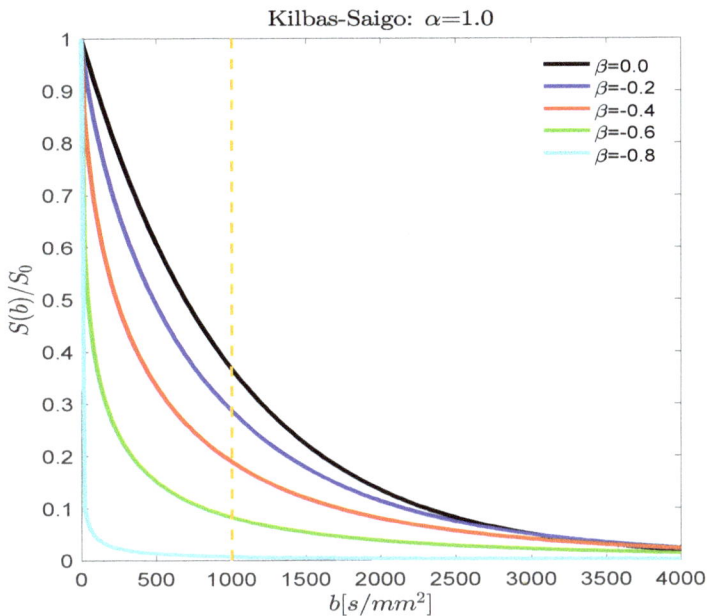

Figure 5. The normalized stretched Kilbas-Saigo signal decay of $S(b)/S_0$ is plotted versus b-values between 0 and 4000 s/mm^2, and $D_0 = 1 \times 10^{-3}$ mm^2/s with $\alpha = 1$, and $\beta = 0, -0.2, -0.4, -0.6$, and -0.8. The signal decay in the $\alpha = 1$ case is a weighted stretched exponential. As the value of β decreases, the shape of the decay curve transitions from a stretched exponential decay to a pattern with a asymptotic power law decay.

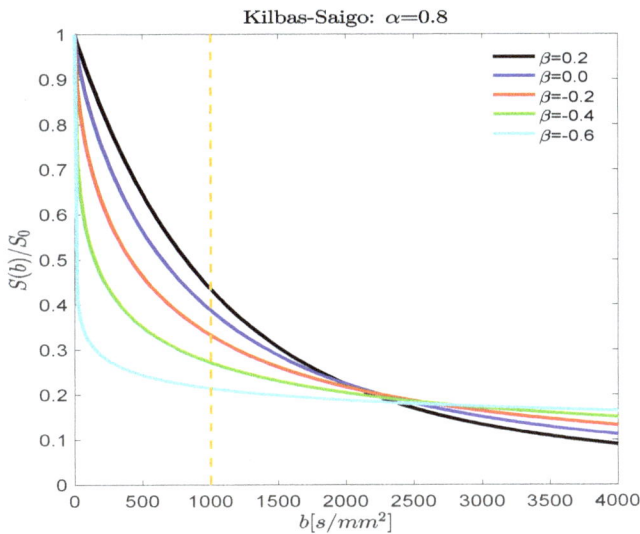

Figure 6. The normalized stretched Kilbas-Saigo signal decay of $S(b)/S_0$ is plotted versus b-values between 0 and 4000 s/mm^2, and $D_0 = 1 \times 10^{-3}$ mm^2/s with $\alpha = 0.8$, and $\beta = 0.2, 0, -0.2, -0.4$, and -0.6. In this figure, we observe the effects of both α and β on the decay.

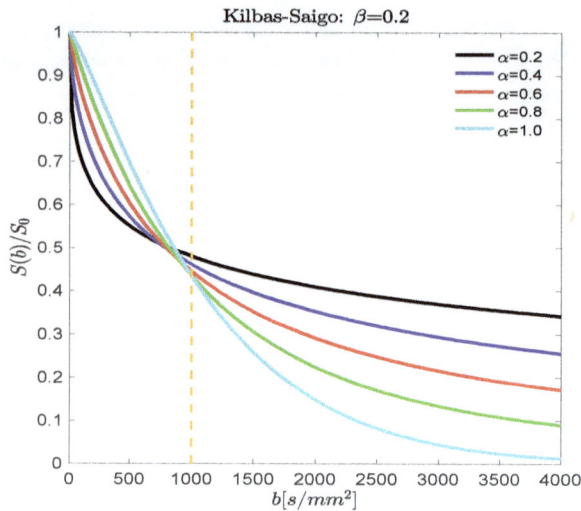

Figure 7. The normalized stretched Kilbas-Saigo signal decay of $S(b)/S_0$ is plotted versus b-values between 0 and 4000 s/mm^2, and $D_0 = 1 \times 10^{-3}$ mm^2/s with $\beta = 0.2$, and $\alpha = 0.2, 0.4, 0.6, 0.8$, and 1.0. In the figure, we see for a fixed power law decay of $D(b)$ the shift from lower to higher decay rates as α approaches the exponential function.

3.2. Fitting Fractional Order D(b) Models to DW-MRI Data

As an example application of these DW-MRI signal decay models, we fit the stretched Kilbas-Saigo function to published ex vivo data acquired from an intact bovine optic nerve [35]. Fits to the exponential, stretched exponential, stretched one parameter Mittag-Leffler were also performed and the results are provided as Supplementary Material. Since the optic nerve consists of bundles of axons, the diffusion signals were measured both along (parallel) and across (perpendicular) the axons. Given the cylindrical structure of the individual nerve axons, we would expect the diffusion to be hindered in the direction along the nerve, and restricted in the direction across the nerve. The Stejskal-Tanner diffusion pulse sequence [5] consists of a pair of rectangular gradient pulses of height g, duration δ, and separation Δ. It was applied by varying the gradient strength, g, and direction for a series of increasing diffusion times ($\Delta = 8, 10, 20$, and 30 ms) with δ set to 3 ms in all cases. The increase in the diffusion time between the gradient pulses can be expected to increase the restricted diffusion contribution for both gradient directions (i.e., when the diffusion length, $L_D = (2D\Delta)^{1/2}$ exceeds the radius of the axons, typically 5 microns in the parallel case and less than one micron in the perpendicular case). The data was extracted from Figure 1 in [35], and fit using nonlinear least squares optimization in MATLAB®. The mean squared error was calculated for the Kilbas-Saigo function in both the parallel and perpendicular gradient directions. These results are presented in Table 2. As expected, the parallel gradient direction exhibited larger values of the diffusion coefficient than the perpendicular gradient direction (0.72×10^{-3} mm^2/s versus 0.2×10^{-3} mm^2/s, when $\Delta = 30$ ms). The stretched Kilbas-Saigo function was able to fit both the parallel and the perpendicular data for all diffusion times. Figure 8 is a plot of the parallel and perpendicular cases when $\Delta = 30$ ms. In the parallel case the α value (ca., 0.75) did not change with increasing diffusion time, while the β value decreased from 0.33 to 0.06. In the perpendicular case, the α value fell from 0.62 to 0.38 with increasing diffusion time, while the β value decreased only from 0.31 to 0.20. These results suggest that the stretched Kilbas-Saigo function is sensitive to both gradient direction and diffusion time. Our goal in this proof of concept paper was only to introduce the Kilbas-Saigo function as a candidate for modeling DW-MRI data. Nevertheless, the nature of the α and β in the model offers the potential of distinguishing between multiexponential (compartmental) contributions via the α parameter and tissue complexity (surface to volume ratio and membrane permeability) contributions through the β parameter. Further analysis of whole imaging slices is needed to establish whether or not this, or another model provides better contrast.

Table 2. Summary of Fitting Results for the Stretched Kilbas-Saigo Decay Model.

		$\Delta = 30$ ms	$\Delta = 20$ ms	$\Delta = 10$ ms	$\Delta = 8$ ms
Parallel	$D \times 104$ mm^2/s	7.19	7.14	8.86	8.74
Perpendicular		1.95	2.05	2.72	2.75
Parallel	α	0.76	0.75	0.64	0.74
Perpendicular		0.38	0.42	0.38	0.62
Parallel	β	0.06	0.14	0.33	0.26
Perpendicular		0.31	0.27	0.46	0.20
Parallel	Mean-squared error	5.84×10^{-4}	5.09×10^{-4}	3.47×10^{-4}	1.64×10^{-4}
Perpendicular		4.38×10^{-4}	2.77×10^{-4}	2.53×10^{-4}	4.61×10^{-4}

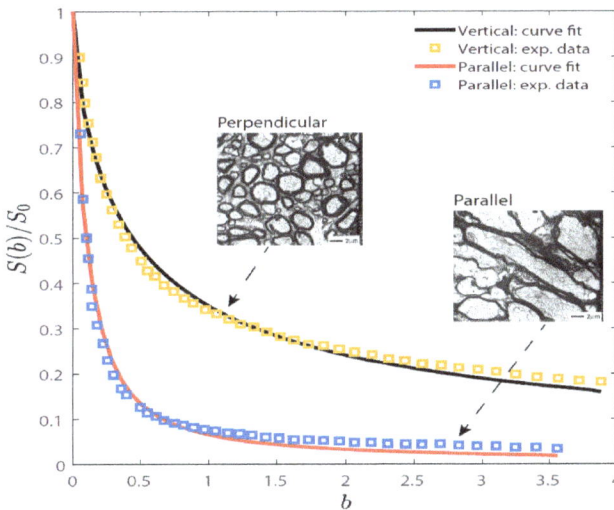

Figure 8. An example of the normalized stretched Kilbas-Saigo signal decay fits of $S(b)/S_0$ to the bovine optic nerve data for the parallel and perpendicular directions with $\Delta = 30$ ms. The photomicrographs are from [35] and are used with permission from the journal Magnetic Resonance in Medicine.

4. Discussion

"One cannot collect all the beautiful shells on the beach" Anne Morrow Lindbergh, *Gift from the Sea*, 1955.

Fractional calculus interpolates between the integer order operations of calculus just as the real numbers are interspersed between integers. Investigators can employ various fractional derivatives (e.g., Grünwald-Letnikov, Riemann-Liouville, Caputo, Riesz, Hadamard, and Erdélyi-Kober) to stitch together earlier or surrounding data in an extended space-time model of a physical process such as diffusion or dielectric relaxation [19]. The fractional derivatives threads are woven into the model via generalized transcendental functions (Mittag-Leffler, Kilbas-Saigo, Wright, and the H-function of Fox) which are functions that often fit the experimental data with greater fidelity and precision [30–32]. In this paper, addressing the decay of diffusion-weighted magnetic resonance signals, we view the problem from a simple perspective—that of a fractional order relaxation with either a fixed or a power-law varying rate constant. The resulting fractional order Kilbas-Saigo model encompasses the simple exponential, the stretched exponential and the stretched Mittag-Leffler function as special cases [21–23].

In previous papers, we investigated the challenge of extracting specific complexity measures from DW-MRI data [12,13]. The choices are myriad, while the data is often sparse. Nevertheless, by tuning the applied MR pulse sequence (e.g., by changing the gradient direction or by increasing the diffusion time between gradient pulses) to match a particular aspect of the tissue under study, savings in time and an economy of interpretation are possible. For example, in diffusion tensor imaging (DTI), one uses a simple exponential decay function, but applies the diffusion pulse gradients in six or more directions to capture the anisotropy of skeletal muscle or brain white matter [10]. In a similar manner, in diffusion kurtosis imaging (DKI), by applying gradients in multiple directions and at multiple b-values, the signal decay encodes a measure of restricted diffusion [16]. In some situations, we acquire diffusion trace data to reduce the contributions of tissue anisotropy [36], but extend the range of b-values to 4000 s/mm^2 or greater to capture aspects of sub-voxel tissue complexity in the fractional order parameters of the Mittag-Leffler function [37,38].

The connection between tissue complexity and the fractional order diffusion decay process is explicit when extremely short diffusion gradient pulses are employed [12,26]. For this situation, the time-space fractional order diffusion equation is known to express—via the Continuous Time Random Walk (CTRW) generalization of Brownian motion—the underlying traps and jumps in the movement of restricted and hindered tissue water. And, while the short gradient pulse duration sequence ($\Delta > \delta$) is often applied in laboratory NMR and animal MRI, the larger gradient coils used in human scanners are not currently able to meet this condition. Also, complete mapping of the diffusion signal {g, Δ, δ} carries a time penalty that precludes exploration of a wide range of values in a typical clinical scanning protocol. Hence, an underlying theme in our work is to optimize imaging time, model selection and full data space characterization.

What are the trade-offs between the respective fractional order models (decay functions) considered in the paper? In general, as the b-value increases beyond 1000 s/mm^2 (e.g., varying g, for fixed Δ and δ) the diffusion regime moves from the domain of hindered diffusion (where diffusion is Gaussian, and the decay is exponential) to one of restricted diffusion (where the diffusion is non-Gaussian, and the decay is often multi-exponential). In the Gaussian case, $S(b) = S_0 exp[-(bD_0)]$, we note the reduction in the apparent diffusion coefficient from its value of 2.4×10^{-3} mm^2/s in cerebral spinal fluid to 0.8×10^{-3} mm^2/s in brain gray matter by a tortuosity factor of 3, so the apparent diffusion coefficient, ADC = D_{CSF}/3. For intermediate b-values (1000–3000 s/mm^2), the stretched exponential decay function, $S(b) = S_0 exp[-(bD_0)^\alpha]$, can be used, but so can a bi-exponential model, a kurtosis model or the varying diffusion coefficient (VDC) model where $D(b) = D_0 exp(-bD_1)$ [14,39]. When the b-values increase above 3000 s/mm^2, the exponential models all rapidly decay while the signal often persists due to trapped tissue water, which exhibits an inverse power law decrease in signal intensity. This behavior can be captured by the VDC model, but it is also naturally encoded in the Mittag-Leffler decay, $S(b) = S_0 E_\alpha[-(bD_0)^\alpha]$, which for low b-values approximates the stretched exponential, and for high b-values decays asymptotically as $(bD_0)^{-\alpha}$. The Kilbas-Saigo generalization of the Mittag-Leffler function is the mother function (model) for all three cases, and has been applied before by Hanyga and Seredynska [34] to diffusion in MRI and by Lin to problems in NMR [40,41]. The behavior of the Kilbas-Saigo modeling approach for the case of varying Δ (for fixed g and δ) has not yet been fully examined. Here, we were successful in capturing the anisotropy of the bovine optic nerve and the emergence of anomalous diffusion as the diffusion time increases. These results are similar to the behavior of the q-space model, $S(b) = S_0 E_{\alpha,\beta}[-D_{\alpha,\beta} q^\beta (\Delta - \delta/3)^\alpha]$, where $q = (\gamma g \delta)/2\pi$ is a measure of the strength of the diffusion gradient phase encoding [26]. Future such studies would involve normal and tumor tissues and could employ oscillating diffusion gradients for short times and stimulated echo imaging for long times.

The motivation for fractional calculus models of diffusion signal decay in MRI has both a practical and a theoretical basis [42]. The practical emerges from advances in diffusion–weighted

Mathematics **2019**, *7*, 348

imaging sequences that use stronger gradients (*b*-values, above 2000 s/mm^2) to probe sub-voxel tissue compartments. The signal from regions with relatively unhindered diffusion (e.g., cerebral spinal fluid) decays very quickly, while the signal from regions exhibiting pronounced hindered and restricted diffusion persists. Researchers and clinicians fit these signals to a variety of models (e.g., multi-exponential, stretched exponential, and kurtosis) to account for the change in the apparent diffusion coefficient as the *b*-value increases—fits that provide additional imaging biomarkers of sub-voxel structure [43–45]. In this ongoing effort, the medical community seeks to link the model parameters to changes in tissue due to local and diffuse disease. The theoretical basis for considering fractional calculus models emerges from the growing recognition in the biophysics community of the prevalence of anomalous diffusion—due to molecular crowding—in cells and cell membranes [8,9]. Mathematical models of anomalous diffusion include Gaussian models (Brownian and fractional Brownian motion), Langevin equations, the non-Gaussian CTRW, and Lorentz obstructed transport models [46,47]. Within this group, there is a recurrence of power law behavior in the statistics of stochastic models, the correlation coefficients of dynamic models and the waiting time and jump increment probabilities in non-Gaussian models. Since the integral and differential operators of fractional calculus naturally accommodate power laws, via convolution, it is not surprising that anomalous diffusion can be cast in the mathematical language of fractional calculus. Alternatively, the diffusion process can be described by generalizing the diffusion constant into a diffusion coefficient that is a function of either *b*-value or diffusion time.

5. Conclusions

In this paper, we described DW-MRI signal decay using sub-diffusion models that can be expressed simply in terms of *b*-value: exponential, stretched exponential, Mittag-Leffler and Kilbas-Saigo. The stretched Kilbas-Saigo function encompasses this set, and depending on the gradient direction and range of *b*-values acquired in the experiment, can fit the data with three parameters: (i) D_0, the apparent diffusion coefficient; (ii) α, the order of the fractional derivative; and (iii) β, the fractional exponent of the inverse power law decay of $D(b)$. The goal was to illustrate how this model can capture the anomalous decay of $S(b)$ observed when data collected over a range of diffusion times. The resulting model parameters encode aspects of the sub-voxel tissue structure, hence offer the potential for describing changes that are brought about by the onset of disease or the success of treatment.

Supplementary Materials: Supplementary materials can be found at http://www.mdpi.com/2227-7390/7/4/348/s1.

Author Contributions: This work involved all coauthors. R.L.M. conceived the idea, wrote the original draft and contributed to the software, the investigation and the editing. H.K. analyzed the functions, wrote the plotting software and contributed to the visualization. S.W. analyzed the data and investigated the results. Y.L. investigated the functional behavior, supervised the data analysis, and edited the manuscript.

Funding: This research received no external funding.

Acknowledgments: The authors would like to acknowledge a discussion with Professor Greg J. Stanisz of the University of Toronto about diffusion in the bovine optic nerve and the work he published in [35]. We would also like to acknowledge the preliminary analysis of the data in Figure 8 of [35] conducted by Davide Carloni, a graduate student in the Department of Bioengineering at UIC.

Conflicts of Interest: The authors declare no conflict of interest.

References

1. Gershenfeld, N. *The Nature of Mathematical Modeling*; Cambridge University Press: Cambridge, UK, 1999; 344p.
2. Haacke, E.M.; Brown, R.W.; Thompson, M.R.; Venkatesan, R. *Magnetic Resonance Imaging: Physical Principles and Sequence Design*; Wiley-Blackwell: New York, NY, USA, 1999; 914p.
3. Jones, D.K. *Diffusion MRI: Theory, Methods, and Applications*; Oxford University Press: Oxford, UK, 2011; 748p.
4. Callaghan, P.T. *Translational Dynamics and Magnetic Resonance: Principles of Pulsed Gradient Spin Echo NMR*; Oxford University Press: Oxford, UK, 2011; 568p.

5. Stejskal, E.O.; Tanner, J.E. Spin diffusion measurements: Spin echoes in the presence of a time-dependent field gradient. *J. Chem. Phys.* **1965**, *42*, 288–292. [CrossRef]
6. Liang, Z.-P.; Lauterbur, P.C. *Principles of Magnetic Resonance Imaging: A Signal Processing Approach*; Wiley—IEEE Press: Piscataway, NJ, USA, 1999; 416p.
7. Kimmich, R. *NMR: Tomography, Diffusometry, Relaxometry*; Springer: New York, NY, USA, 1997; 526p.
8. Bouchaud, J.-P.; Georges, A. Anomalous diffusion in disordered media: Statistical mechanisms, models and physical applications. *Phys. Rep.* **1990**, *195*, 127–293. [CrossRef]
9. Höfling, F.; Franosch, T. Anomalous transport in the crowded world of biological cells. *Rep. Prog. Phys.* **2013**, *76*, 046602. [CrossRef] [PubMed]
10. Mori, S.; Tournier, J.D. *Introduction to Diffusion Tensor Imaging: And Higher Order Models*; Academic Press: New York, NY, USA, 2013; 126p.
11. Price, W.S. *NMR Studies of Translational Motion: Principles and Applications*; Cambridge University Press: Cambridge, UK, 2009; 393p.
12. Magin, R.L.; Ingo, C.; Colon-Perez, L.; Triplett, W.; Mareci, T.H. Characterization of anomalous diffusion in porous biological tissues using fractional order derivatives and entropy. *Microporous Mesoporous Mater.* **2013**, *178*, 39–43. [CrossRef]
13. Magin, R.L.; Abdullah, O.; Baleanu, D.; Zhou, X.J. Anomalous diffusion expressed through fractional order differential operators in the Bloch-Torrey equation. *J. Magn. Reson.* **2008**, *190*, 255–270. [CrossRef] [PubMed]
14. Magin, R.L. Models of diffusion signal decay in magnetic resonance imaging: Capturing complexity. *Concepts Magn. Reson.* **2016**, *45A*, 21401. [CrossRef]
15. Alexander, D.C.; Hubbard, P.L.; Hall, M.G.; Moore, E.A.; Ptito, M.; Parker, G.J.M.; Dyrby, T.B. Orientationally invariant indices of axon diameter and density from diffusion MRI. *Neuroimage* **2010**, *52*, 1374–1389. [CrossRef] [PubMed]
16. Jensen, J.H.; Helpern, J.A.; Ramani, A.; Lu, H.; Kaczynski, K. Diffusional kurtosis imaging: The quantification of non-Gaussian water diffusion by means of magnetic resonance imaging. *Magn. Reson. Med.* **2005**, *53*, 1432–1440. [CrossRef]
17. Bennett, K.M.; Schmaidna, K.M.; Bennett, R.; Rowe, D.B.; Lu, H.; Hyde, J.S. Characterization of continuously distributed cortical water diffusion rates with a stretched-exponential model. *Magn. Reson. Med.* **2003**, *50*, 727–734. [CrossRef]
18. Jelescu, I.O.; Veraart, J.; Fieremans, E.; Novikov, D.S. Degeneracy in model parameter estimation for multi-compartmental diffusion in neuronal tissue. *NMR Biomed.* **2016**, *29*, 33–47. [CrossRef] [PubMed]
19. Podlubny, I. *Fractional Differential Equations*; Academic Press: New York, NY, USA, 1999; 368p.
20. Gorenflo, R.; Kilbas, A.A.; Mainardi, F.; Rogosin, S.V. *Mittag-Leffler Functions, Related Topics and Applications*; Springer-Verlag: Berlin, Germany, 2014; 443p.
21. Kilbas, A.A.; Saigo, M.; Saxena, R.K. Generalized Mittag-Leffler function and generalized fractional calculus operators. *Integr. Trans. Spec. Funct.* **2004**, *15*, 31–49. [CrossRef]
22. de Oliveira, E.C.; Mainardi, F.; Vaz, J. Fractional models of anomalous relaxation based on the Kilbas and Saigo function. *Meccanica* **2014**, *49*, 2049–2060. [CrossRef]
23. Kilbas, A.A.; Srivastava, H.M.; Trujillo, J.J. *Theory and Applications of Fractional Differential Equations*; Elsevier: Amsterdam, The Netherlands, 2006; 523p.
24. Hall, M.G.; Barrick, T.R. From diffusion-weighted MRI to anomalous diffusion imaging. *Magn. Reson. Med.* **2008**, *59*, 447–455. [CrossRef]
25. Berberan-Santos, M.N. A luminescence decay function encompassing the stretched exponential and the compressed hyperbola. *Chem. Phys. Lett.* **2008**, *460*, 146–150. [CrossRef]
26. Ingo, C.; Magin, R.L.; Colon-Perez, L.; Triplett, W.; Mareci, T.H. On random walks and entropy in diffusion weighted magnetic resonance imaging studies of neural tissue. *Magn. Reson. Med.* **2014**, *71*, 617–627. [CrossRef]
27. R.L.M. Kilbas and Saigo Function. Available online: https://www.mathworks.com/matlabcentral/fileexchange/70999-kilbas-and-saigo-function (accessed on 12 March 2018).
28. Berberan-Santos, M.N.; Bodunov, E.N.; Valeur, B. Mathematical functions for the analysis of luminescence decays with underlying distributions: 1. Kohlrausch decay function (stretched exponential). *Chem. Phys.* **2005**, *315*, 171–182. [CrossRef]

29. Berberan-Santos, M.N.; Valeur, B. Luminescence decays with underlying distributions: General properties and analysis with mathematical functions. *J. Lumin.* **2007**, *126*, 263–272. [CrossRef]

30. Spanier, J.; Oldham, K.B. *An Atlas of Functions*; Hemisphere Publishing Corp: New York, NY, USA, 1987; pp. 253–262.

31. Carpinteri, A.; Mainardi, F. *Fractals and Fractional Calculus in Continuum Mechanics CISM Courses and Lectures No. 378*; Springer-Wien: New York, NY, USA, 1997; 348p.

32. Hilfer, R. (Ed.) *Applications of Fractional Calculus in Physics*; World Scientific: Singapore, 2007; 472p.

33. Berberan-Santos, M.N. Relation between the inverse Laplace transforms of I(tb) and I(t): Application to the Mittag-Leffler and asymptotic inverse power law relaxation functions. *J. Math. Chem.* **2005**, *38*, 265–270. [CrossRef]

34. Hanyga, A.; Seredynska, M. Anisotropy in high-resolution diffusion-weighted MRI and anomalous diffusion. *J. Magn. Reson.* **2012**, *220*, 85–93. [CrossRef]

35. Stanisz, G.J.; Szafer, A.; Wright, G.A.; Henkelman, R.M. An analytical model of restricted diffusion in bovine optic nerve. *Magn. Reson. Med.* **1997**, *37*, 103–111. [CrossRef]

36. Zhou, X.J.; Gao, Q.; Abdullah, O.; Magin, R.L. Studies of anomalous diffusion in the human brain using fractional order calculus. *Magn. Reson. Med.* **2010**, *63*, 562–569. [CrossRef]

37. Sui, Y.; He, W.; Damen, F.W.; Wanamaker, C.; Li, Y.; Zhou, X.J. Differentiation of low- and high-grade pediatric brain tumors with high b-value diffusion weighted MR imaging and a fractional order calculus model. *Radiology* **2015**, *277*, 489–496. [CrossRef]

38. Karaman, M.M.; Sui, Y.; Wang, H.; Magin, R.L.; Li, Y.; Zhou, X.J. Differentiating low- and high-grade pediatric brain tumors using a continuous-time random-walk diffusion model at high b-values. *Magn. Reson. Med.* **2016**, *76*, 1149–1157. [CrossRef]

39. Magin, R.L.; Karaman, M.M.; Hall, M.G.; Zhu, W.; Zhou, X.J. Capturing complexity of the diffusion-weighted MR signal decay. *Magn. Reson. Imaging* **2019**, *56*, 110–118. [CrossRef]

40. Lin, G. Analysis of PFG anomalous diffusion via real-space and phase-space approaches. *Mathematics* **2018**, *6*, 17. [CrossRef]

41. Lin, G. Describe NMR relaxation by anomalous rotational or translational diffusion. *Commun. Nonlinear Sci. Numer. Simul.* **2019**, *72*, 232–239. [CrossRef]

42. Novikov, D.S.; Kiselev, V.G.; Jespersen, S.N. On modeling. *Magn. Reson. Med.* **2018**, *79*, 317–3193. [CrossRef]

43. Grebenkov, D.S. From the microstructure to diffusion NMR, and back. In *Diffusion NMR of Confined Systems: Fluid Transport in Porous Solids and Heterogeneous Materials*; Valiullin, R., Ed.; The Royal Society of Chemistry: Cambridge, UK, 2017; pp. 52–110.

44. Grinberg, F.; Farrher, F.; Shah, N.J. Diffusion magnetic resonance imaging in brain tissue. In *Diffusion NMR of Confined Systems: Fluid Transport in Porous Solids and Heterogeneous Materials*; Valiullin, R., Ed.; The Royal Society of Chemistry: Cambridge, UK, 2017; pp. 497–528.

45. Topgaard, D. Multidimensional diffusion MRI. *J. Magn. Reson.* **2019**, *275*, 98–113. [CrossRef]

46. Klages, R.; Radons, G.; Sokolov, I.M. (Eds.) *Anomalous Transport: Foundations and Applications*; Wiley-VCH: Weinheim, Germany, 2008; 583p.

47. Evangelista, L.R.; Lenzi, E.K. *Fractional Diffusion Equations and Anomalous Diffusion*; Cambridge Univ. Press: Cambridge, UK, 2018; 345p.

mathematics

Article

Adaptive Pinning Synchronization of Fractional Complex Networks with Impulses and Reaction–Diffusion Terms

Xudong Hai, Guojian Ren, Yongguang Yu * and Conghui Xu

Department of Mathematics, Beijing Jiaotong University, Beijing 100044, China; 18118003@bjtu.edu.cn (X.H.); 15118418@bjtu.edu.cn (G.R.); 17118433@bjtu.edu.cn (C.X.)
* Correspondence: ygyu@bjtu.edu.cn

Received: 7 March 2019; Accepted: 24 April 2019; Published: 7 May 2019

Abstract: In this paper, a class of fractional complex networks with impulses and reaction–diffusion terms is introduced and studied. Meanwhile, a class of more general network structures is considered, which consists of an instant communication topology and a delayed communication topology. Based on the Lyapunov method and linear matrix inequality techniques, some sufficient criteria are obtained, ensuring adaptive pinning synchronization of the network under a designed adaptive control strategy. In addition, a pinning scheme is proposed, which shows that the nodes with delayed communication are good candidates for applying controllers. Finally, a numerical example is given to verify the validity of the main results.

Keywords: fractional complex networks; adaptive control; pinning synchronization; time-varying delays; impulses; reaction–diffusion terms

---

1. Introduction

Complex networks, which are composed of a large set of nodes connected by edges, are used to describe many large-scale systems in nature and human societies, such as the Internet, power grid networks, the World Wide Web and so on [1–3]. Therefore, it is very meaningful and important to investigate the dynamical behaviors of complex networks. Up until now, there have been a large number of excellent scientific research results on the dynamical analysis of complex networks, including of synchronization [4], state estimation [5], impulsive control [6], and so on.

Synchronization, as one of the most important collective behaviors of complex dynamical systems, widely exists in the world, ranging from natural systems to man-made networks [7,8]. In recent years, synchronization has received a great deal of attention due to its potential application in various fields, including signal processing, secure communication, biological systems, and so on [9–11]. Various types of synchronization problems have been studied, such as projective lag synchronization [12], cluster synchronization [13], and exponential synchronization [14]. However, as we know, it is impossible for most dynamical systems to achieve synchronization by themselves in many real situations. Some control strategies must be designed to force the systems to be synchronized, such as feedback control [15], impulsive control [16], and adaptive control [17]. It could be a waste of money and resources op add controllers to all nodes in a large-scale network. In order to solve this problem, a pinning control strategy, which involves only a small fraction of all the nodes being controlled to force networks to become synchronized, has been proposed and extensively studied by researchers [18,19]. To reduce the enormous difference in control strength between theoretical values and practical needs, adaptive control, as a valid method, has been discussed in literature [20,21].

In practice, diffusion phenomena cannot be ignored. For example, electrons move in a nonuniform electromagnetic field and, in the process of chemical reactions, different chemicals react with each other and spatially diffuse in the inter-medium until a balanced-state spatial concentration pattern has been structured. Thus, it is reasonable to consider complex networks with reaction–diffusion terms. In recent years, some papers concerning the control and synchronization of complex dynamical systems with diffusion effects terms have been published [22–24]. In the real world, the state of artificial and biological networks are often subject to instantaneous perturbations and experience abrupt changes at certain instants, which may be caused by switching phenomena, frequency changes, or other sudden noises. These instantaneous perturbations always exhibit impulsive effects. Impulsive dynamical systems have naturally received a lot of attention, and some excellent results have been published [25,26]. Therefore, it is necessary to investigate the influence of impulsive perturbations on complex networks.

It should be noted that all of the mathematical models in this paper are fractional systems. As a generalization of traditional calculus, fractional calculus [27] provides an excellent instrument for the description of memory and the hereditary properties of various materials and processes. With the development of scientific research, these advantageous properties of fractional integration and differentiation have been noticed by more and more researchers, and fractional calculus, as a useful tool, has been widely applied in many fields such as mathematical biology [28], signal processing [29], and so on.

In addition, to be more practical, it is necessary to consider time delay. Due to the finite switching speed of amplifiers and the finite signal propagation time, time delay naturally exists in all kinds of dynamical systems and is unavoidable, which may lead to instability, chaos, or other performances of dynamical systems [30,31]. Thus, it is valuable to investigate the phenomenon of time delay in complex dynamical systems, and some outstanding results from research into time delay have been published, such as [32].

Motivated by the above discussions, the main contributions of this paper are as follows. (1) A class of fractional complex networks with impulses and reaction–diffusion terms is studied. Meanwhile, a class of more general network structures is considered in which all nodes are divided into three categories: nodes can only send information to others instantly; nodes can only be connected with others with a time delay; and nodes can communicate both instantly and with a delay. (2) Some sufficient conditions are derived to ensure the adaptive pinning synchronization of the network. (3) A pinning scheme is designed that shows that some delay-coupled nodes should be pinned first.

The rest of this paper is arranged as follows: some preliminaries and the model description of fractional networks with impulses and reaction–diffusion terms are provided in Section 2; in Section 3, the main results of this paper are described; next, in Section 4, a numerical simulation is presented to illustrate the effectiveness and correctness of the main results; finally, the conclusion of this paper is given in Section 5.

Notations: Let $R^+ = [0, +\infty)$, $R = (-\infty, +\infty)$, R^n be the n-dimensional Euclidean space, and $R^{n \times m}$ be the space of $n \times m$ real matrices. $P \in R^{n \times n} \geq 0$ ($P \in R^{n \times n} \leq 0$) means that matrix P is symmetric and semi-positive (semi-negative) definite. $P \in R^{n \times n} > 0$ ($P \in R^{n \times n} < 0$) means that matrix P is symmetric and positive (negative) definite. I_n denotes an $n \times n$ real identity matrix. A^T is the transpose of matrix A. B^{-1} means the inverse of matrix B. Let $A^2 = A^T A$ where $A \in R^{n \times n}$. $\otimes$ represents the Kronecker product of two matrices. $\lambda_{min}(\cdot)$ and $\lambda_{max}(\cdot)$ denote the minimum and the maximum eigenvalue of the corresponding matrix, respectively. $\| \cdot \|$ denotes the Euclidean norm of a vector, for example, $\| \beta \| = \sqrt{\sum_{i=1}^{n} \beta_i^2}$ where $\beta = (\beta_1, \cdots, \beta_n)^T$.

2. Preliminaries and System Description

In this section, some basic definitions of fractional calculus are introduced as the preliminaries of this paper, and some necessary conclusions are presented for use in the next several sections. Meanwhile, the mathematical model of a class of fractional complex networks with impulses and reaction–diffusion terms is described and the definition of synchronization of complex networks is given.

2.1. Fractional Integral and Derivative

Definition 1 ([33]). *Riemann–Liouville fractional derivative with order α for a function $x : R^+ \to R$ is defined as*

$$_{t_0}^R D_t^\alpha x(t) = \frac{1}{\Gamma(m-\alpha)} \frac{d^m}{dt^m} \int_{t_0}^t (t-\tau)^{m-\alpha-1} x(\tau) d\tau,$$

where $0 \le m-1 \le \alpha < m$, $m \in Z_+$, and Z_+ denotes the collection of all positive integers. $\Gamma(\cdot)$ is the gamma function.

Definition 2 ([33]). *Riemann–Liouville fractional integral of order α for a function $f : R^+ \to R$ is defined by*

$$_{t_0}^R I_t^\alpha f(t) = \frac{1}{\Gamma(\alpha)} \int_{t_0}^t (t-\tau)^{\alpha-1} f(\tau) d\tau,$$

where $\alpha > 0$ and $\Gamma(\cdot)$ is the gamma function.

Lemma 1 ([33]). *For any constants k_1 and k_2, the linearity of the Riemann–Liouville fractional derivative is described by*

$$_{t_0}^R D_t^\alpha (k_1 f(t) + k_2 g(t)) = k_1 {}_{t_0}^R D_t^\alpha f(t) + k_2 {}_{t_0}^R D_t^\alpha g(t).$$

Lemma 2 ([33]). *If $p > q > 0$, then the following equality*

$$_{t_0}^R D_t^p {}_{t_0}^R I_t^q f(t) = {}_{t_0}^R D_t^{p-q} f(t)$$

holds for sufficiently good functions $f(t)$. In particular, this relation holds if $f(t)$ is integrable.

Lemma 3 ([34]). *Let $x(t) : R^n \to R^n$ be a vector of differentiable function. Then, for any time instant $t \ge t_0$, the following inequality holds*

$$_{t_0}^R D_t^\alpha (x^T(t) P x(t)) \le 2 x^T(t) (P {}_{t_0}^R D_t^\alpha x(t)),$$

where $0 < \alpha < 1$ and $P \in R^{n \times n}$ is a constant, square, symmetric, and positive definite matrix.

Lemma 4 ([35]). *Let Ω be a cube $|x_k| < l_k$ $k = 1, 2, \cdots, q$, and let $h(x)$ be a real-valued function belonging to $C^1(\Omega)$, which vanishes on the boundary $\partial\Omega$ of Ω, i.e., $h(x)|_{\partial\Omega} = 0$. Then,*

$$\int_\Omega h^2(x) dx \le l_k^2 \int_\Omega (\frac{\partial h}{\partial x_k})^2 dx,$$

where $x = (x_1, x_2, \cdots, x_q)^T$.

Lemma 5 ([36]). *The following linear matrix inequality (LMI)*

$$L = \begin{bmatrix} Q(x) & S(x) \\ S(x)^T & R(x) \end{bmatrix} < 0,$$

where $Q(x)^T = Q(x)$ and $R(x)^T = R(x)$, is equivalent to any one of the following conditions:

(1) $Q(x) < 0$, $R(x) - S(x)^T Q(x)^{-1} S(x) < 0$;

(2) $R(x) < 0$, $Q(x) - S(x) R(x)^{-1} S(x)^T < 0$.

Lemma 6 ([37]). *For any $x, y \in R^n$, $\epsilon > 0$, the inequality $2x^T y \le \epsilon x^T x + \frac{1}{\epsilon} y^T y$ holds.*

2.2. Theories of Graphs and Matrices

An undirected graph $\mathcal{G} = \{\mathcal{V}, \mathcal{E}\}$ consists of nodes $\mathcal{V} = \{1, 2, \cdots, N\}$ and a set of edges $\mathcal{E}$. If there is an edge between nodes i and j, then $(i, j) \in \mathcal{E}$, and node j is called the neighbor of node i and vice versa. The Laplacian matrix $L = (L_{ij})_{N \times N}$ representing the topological structure of graph $\mathcal{G}$ is defined as: when $i \neq j$, $L_{ij} = L_{ji} < 0$ if $(i, j) \in \mathcal{E}$; otherwise, $L_{ij} = L_{ji} = 0$; $L_{ii} = -\sum_{j=1, j \neq i}^{N} L_{ij} = -\sum_{j=1, j \neq i}^{N} L_{ji}$, which ensures the property that $\sum_{j=1}^{N} L_{ij} = \sum_{i=1}^{N} L_{ij} = 0$.

Lemma 7 ([38]). *If L is the Laplacian matrix of a connected network, $L_{ij} = L_{ji} \leq 0$ for $i \neq j$, and $\sum_{j=1}^{N} L_{ij} = 0$ for all $i = 1, \ldots, N$, then all eigenvalues of the matrix*

$$
\begin{bmatrix}
L_{11} + \varepsilon & L_{12} & \cdots & L_{1N} \\
L_{21} & L_{22} & \cdots & L_{2N} \\
\vdots & \vdots & \ddots & \vdots \\
L_{N1} & L_{N2} & \cdots & L_{NN}
\end{bmatrix}
$$

are positive for any positive constant ε.

Lemma 8 ([39]). *For matrices A, B, C, and D with appropriate dimensions, the Kronecker product $\otimes$ satisfies*

(1) $(\theta A) \otimes B = A \otimes (\theta B)$, *where θ is a constant;*
(2) $(A + B) \otimes C = A \otimes C + B \otimes C$;
(3) $(A \otimes B)(C \otimes D) = (AC) \otimes (BD)$;
(4) $(A \otimes B)^T = A^T \otimes B^T$.

Lemma 9 ([40]). *The Laplacian matrix L in an undirected graph is semi-positive definite. It has a simple zero eigenvalue, and all the other eigenvalues are positive if and only if the graph is connected.*

2.3. System Description

In this paper, a class of fractional complex networks with time-varying delays is considered. It is assumed that there exist two different modes of communication between the nodes in a network: instant communication and delayed communication. Namely, the network structure consists of two topologies, the instant communication topology $\mathcal{G}$ and the delayed communication topology $\hat{\mathcal{G}}$. $\mathcal{G} = \{\mathcal{V}, \mathcal{E}\}$ and $\hat{\mathcal{G}} = \{\mathcal{V}, \hat{\mathcal{E}}\}$, where $\mathcal{E}$ and $\hat{\mathcal{E}}$ denote the sets of instant communication links and delayed communication links, respectively. Let $L = (L_{ij})_{N \times N}$ and $\hat{L} = (\hat{L}_{ij})_{N \times N}$ represent the Laplacian matrices of $\mathcal{G}$ and $\hat{\mathcal{G}}$, respectively. It is noted that each node in the network has at least one mode of communication and that the graphs $\mathcal{G}$ and $\hat{\mathcal{G}}$ can be disconnected, which implies that they consist of several connected components, and each component is considered as a cluster (group or family). In other words, if there are p nodes that can only communicate with other nodes instantly and q nodes that can only be connected to others with delays, then there exist two elementary matrices F and H such that

$$
F^T L F = \begin{bmatrix}
L_1 & 0 & 0 & 0 & 0 \\
0 & L_2 & 0 & 0 & 0 \\
\vdots & & \ddots & \vdots & \vdots & \vdots \\
0 & 0 & 0 & L_m & 0 \\
0 & 0 & 0 & 0 & 0_{q \times q}
\end{bmatrix}, \quad
H^T \hat{L} H = \begin{bmatrix}
\hat{L}_1 & 0 & 0 & 0 & 0 \\
0 & \hat{L}_2 & 0 & 0 & 0 \\
\vdots & & \ddots & \vdots & \vdots & \vdots \\
0 & 0 & 0 & \hat{L}_l & 0 \\
0 & 0 & 0 & 0 & 0_{p \times p}
\end{bmatrix}, \tag{1}
$$

where L_i and $\hat{L}_i$ are the Laplacian matrices of the ith connected component of $\mathcal{G}$ and $\hat{\mathcal{G}}$, respectively. Then the mathematical model of fractional complex networks with impulses and reaction–diffusion terms is considered as follows:

$$\begin{cases} \dfrac{\partial^\alpha z_i(x,t)}{\partial t^\alpha} = D\Delta z_i(x,t) + f(z_i(x,t)) - c\sum_{j=1}^{N} L_{ij}\Gamma z_j(x,t) - \hat{c}\sum_{j=1}^{N} \hat{L}_{ij}\hat{\Gamma} z_j(x,t-\tau(t)), i = 1,\cdots,N, t \neq t_k; \\ z_i(x,t_k^+) = z_i(x,t_k^-) + p_{ik}(x,t_k), \ k = 1,2,\cdots, \end{cases} \quad (2)$$

where ∂^α denotes the Riemann–Liouville fractional derivative operator of order α ($0 < \alpha < 1$); $x = (x_1, x_2, \ldots, x_q)^T \in \Omega \subset R^q$, $\Delta = \sum_{k=1}^{q}(\frac{\partial^2}{\partial x_k^2})$ is the Laplace diffusion operator on Ω; $z_i(x,t) = (z_{i1}(x,t), z_{i2}(x,t), \ldots, z_{in}(x,t))^T \in R^n$ is the state vector of the ith node at time t and in space x; $D = diag(d_1, d_2, \cdots, d_n) \geq 0$, and d_i is the transmission diffusion coefficient along the ith node; $\tau(t) > 0$ corresponds to the time-varying delay at time t, and it is a continuous function satisfying $0 < \tau(t) \leq \tau^M$, $\dot{\tau}(t) \leq \tau^D < 1$, where $\tau^M = \sup_{t \geq t_0} \tau(t)$ and τ^D are nonnegative constants; $f : R^n \to R^n$ is a continuously differentiable vector function; $c \geq 0$ and $\hat{c} \geq 0$ are the coupling strength; $\Gamma = diag(\gamma_1, \gamma_1, \cdots, \gamma_n) \in R^{n \times n}$ and $\hat{\Gamma} = diag(\hat{\gamma}_1, \hat{\gamma}_1, \cdots, \hat{\gamma}_n) \in R^{n \times n}$ are positive semi-definite inner coupling matrices where $\gamma_j, \hat{\gamma}_j > 0$ if two nodes can communicate through the jth state and $\gamma_j, \hat{\gamma}_j = 0$ otherwise; $L = (L_{ij})_{N \times N}$ and $\hat{L} = (\hat{L}_{ij})_{N \times N}$ are the Laplacian matrices representing the topological structure of the network; t_k ($k = 0, 1, 2, \cdots$) denote impulsive moments and satisfy $t_0 < t_1 < t_2 < \cdots$, $\lim_{k \to \infty} t_k = \infty$, and $t_{k+1} - t_k \geq \omega > 0$; $z_i(x, t_k^-)$ and $z_i(x, t_k^+)$ are the state variables of the ith node before and after impulsive perturbation, respectively; $z_i(x, t_k^-) = \lim_{t \to t_k^-} z_i(x,t)$, $z_i(x, t_k^+) = \lim_{t \to t_k^+} z_i(x,t)$, and assuming that $z_i(x, t_k^-) = z_i(x, t_k)$, the solution of system (2) is left continuous at time t_k; $p_{ik}(x, t_k) : R^n \to R^n$ is the function of the change of $z_i(x,t)$ at time t_k in space x.

Next, the dynamics of an isolated node can be described by

$$\frac{\partial^\alpha z_0(x,t)}{\partial t^\alpha} = D\Delta z_0(x,t) + f(z_0(x,t)), \quad (3)$$

where $z_0(x,t) \in R^n$ is the state vector of the isolated node and may be an equilibrium point, a periodic orbit, or even a chaotic orbit.

Synchronization errors between the relative state of nodes in network and the state of an isolated node can be defined by $e_i(x,t) = z_i(x,t) - z_0(x,t)$, $i = 1, 2, \cdots, N$. Next, the definition of synchronization of complex networks is given as follows

Definition 3. *A complex network* (2) *is said to be synchronized if for any initial condition, the following equality is satisfied:*

$$\lim_{t \to \infty} \| e_i(x,t) \| = \lim_{t \to \infty} \| z_i(x,t) - z_0(x,t) \| = 0, \quad i = 1, 2, \cdots, N.$$

Remark 1. *The model* (2) *has been proposed and studied in [41,42]. But it is different from [41,42] in that impulsive perturbations, reaction–diffusion terms, and a class of more general network structures are considered in this paper. All nodes in the network can be divided into three categories: (1) nodes that can only send information to others instantly; (2) nodes that can only be connected to others with a time delay; and (3) nodes that can communicate with others both instantly and with a delay. Note that the assumption that the whole network should be connected is no longer needed. Thus, the mathematical model is more practical.*

Remark 2. *The impulses in the system* (2) *can be understood as a special property of the system itself or some uncertain disturbances caused by external noises. The pulse period may be no longer fixed or regular, and the intensity of the impulses can be uncertain.*

3. Main Results

In this section, some sufficient criteria for pinning and adaptive synchronization of complex networks (2) are derived, and a pinning scheme is given to discuss which node should be selected first.

Throughout this paper, the following assumptions are needed:

Assumption 1. *The set Ω is that $\Omega = \{x : x = (x_1, x_2, \cdots, x_q)^T, \quad |x_k| < l_k, \ k = 1, 2, \cdots, q\}$ where l_k ($k = 1, 2, \cdots, q$) are positive constants.*

Assumption 2. *There exist functions $\delta_{ik}(x, t_k)$ such that the functions $p_{ik}(x, t_k)$ satisfy*

$$p_{ik}(x, t_k) = -\delta_{ik}(x, t_k)[\sum_{j=1}^{N} L_{ij} z_j(x, t_k)], \quad 0 \le \delta_{ik}(x, t_k) \le \frac{2}{\lambda_{max}(L)},$$

where $i = 1, 2, \cdots, N$, $k = 0, 1, 2, \cdots$. $\delta_{ik}(x, t_k) : R^{q+1} \to R$ stands for the impulsive strength of the ith node at time instants t_k and in space x.

Assumption 3. *(One-sided Lipschitz condition) There exists a constant diagonal matrix $\Xi = diag(\Xi_1, \Xi_2, \ldots, \Xi_n)$ such that*

$$(u(x, t) - v(x, t))^T (f(u(x, t)) - f(v(x, t))) \le (u(x, t) - v(x, t))^T \Xi (u(x, t) - v(x, t)),$$
$$\forall u, v \in R^n, \ t \in R^+, \ x \in \Omega.$$

Remark 3. *Note that Assumption 3 is very mild [43]. For example, all linear and piece-wise linear functions satisfy this condition. In addition, if $\partial f_i / \partial u_j$ ($i, j = 1, 2, \ldots, n$) are bounded, the above condition is satisfied in many well-known systems such as the Lorenz system, Chen system, Lü system, recurrent neural networks, Chua's circuit, and so on. Generally speaking, any function in the system that satisfyies the Lipschitz condition can guarantee this assumption. According to [44], any Lipschitz function is a one-sided Lipschitz function, but the converse is not true. This means that for some nonlinear functions with a big Lipschitz constant, the corresponding one-sided Lipschitz constant matrix Ξ may be a non-positive definite matrix (see the example given in [45]).*

3.1. Adaptive and Pinning Control

The boundary and initial value conditions of complex networks are given in the following form:

$$z_i(x, t) = 0, \quad t \in [t_0 - \tau^M, \infty), \ x \in \partial \Omega; \tag{4}$$

$$z_i(x, s) = \varphi_{0i}(x, s) \in R^n, \quad s \in [t_0 - \tau^M, 0), \ x \in \Omega, \tag{5}$$

where $i = 0, 1, 2, \cdots, N$, and $\varphi_{0i}(x, s)$ is bounded and continuous on $\Omega \times [t_0 - \tau^M, 0)$.

Then, the synchronization error system between systems (2) and (3) is given as follows:

$$\begin{cases} \dfrac{\partial^\alpha e_i(x, t)}{\partial t^\alpha} = D \Delta e_i(x, t) + f(z_i(x, t)) - f(z_0(x, t)) - c \sum_{j=1}^{N} L_{ij} \Gamma e_j(x, t) \\[2mm] \qquad - \hat{c} \sum_{j=1}^{N} \hat{L}_{ij} \hat{\Gamma} e_j(x, t - \tau(t)) + u_i(x, t), \ t \neq t_k; \\[2mm] e_i(x, t_k^+) = e_i(x, t_k^-) - \delta_{ik}(x, t_k)[\sum_{j=1}^{N} L_{ij} e_j(x, t_k)], \quad i = 1, 2, \cdots, N, \ k = 1, 2, \cdots, \end{cases} \tag{6}$$

where $\sum_{j=1}^{N} L_{ij} = 0$ and $\sum_{j=1}^{N} \hat{L}_{ij} = 0$ are used above. $u_i(x, t)$ is the n-dimensional linear feedback controller on the ith node. Without loss of generality, it is assumed that the first l nodes are controlled in the network, and the pinning controller is designed by

$$u_i(x,t) = \begin{cases} -b_i \Gamma e_i(x,t) & i = 1, \ldots, l, \\ 0 & i = l+1, \ldots, N, \end{cases} \tag{7}$$

where $b_i \geq 0$ $(i = 1, 2, \cdots, l)$ denotes control gain.

Next, a theorem is established to derive the synchronization criteria for network (2).

Theorem 1. *Suppose Assumptions 1–3 hold. Under control law* (7), *network* (2) *with initial conditions* (4) *and* (5) *can be synchronized if there exist a positive definite matrix $Q \in R^{n \times n}$ and a positive constant ϵ such that the following inequalities are satisfied:*

$$\begin{bmatrix} \lambda_{max}\left(-\xi D + \Xi + \frac{1}{1-\tau^D}Q\right)I_N - \lambda_{min}(\Gamma)(cL+B) & \frac{\sqrt{2\epsilon\hat{c}}\lambda_{max}(\hat{\Gamma})}{2}\hat{L} \\ \frac{\sqrt{2\epsilon\hat{c}}\lambda_{max}(\hat{\Gamma})}{2}\hat{L} & -I_N \end{bmatrix} < 0, \tag{8}$$

$$\frac{\hat{c}}{2\epsilon}I_n - Q \leq 0, \tag{9}$$

where $\xi = \sum_{k=1}^{q} \frac{1}{l_k^2}$, $B = diag(b_1, b_2, \cdots, b_N)$ and $b_i \geq 0$ $(i = 1, \cdots, l)$, $b_i = 0$ $(i = l+1, \cdots, N)$.

Proof of Theorem 1. Let $\quad \| \quad e(x,t) \quad \|^2 = \quad \sum_{i=1}^{N} e_i^T(x,t)e_i(x,t) \quad$ where $\quad e(t) \quad =$ $(e_1^T(x,t), e_2^T(x,t), \cdots, e_N^T(x,t))^T$. Then, the proof of this theorem can be given in two steps.

Firstly, the case of $t > t_0$ and $t = t_k$, $(k = 1, 2, \cdots)$ is considered. According to Assumption 2, the following inequality holds:

$$\| e(x, t_k^+) \|^2 = \sum_{i=1}^{N} e_i^T(x, t_k^+)e_i(x, t_k^+)$$

$$= \sum_{i=1}^{N} \{e_i(x,t_k^-) - \delta_{ik}(x,t_k)[\sum_{j=1}^{N} L_{ij}e_j(x,t_k)]\}^T \{e_i(x,t_k^-) - \delta_{ik}(x,t_k)[\sum_{j=1}^{N} L_{ij}e_j(x,t_k)]\} \tag{10}$$

$$= e^T(x,t_k)[I_{nN} - (\delta_{x,t_k}L) \otimes I_n]^2 e(x,t_k)$$

$$\leq \| e(x,t_k) \|^2,$$

where $\delta_{x,t_k} = diag(\delta_{1k}(x,t_k), \delta_{2k}(x,t_k), \cdots, \delta_{Nk}(x,t_k))$.

Secondly, the case of $t \geq t_0$ and $t \in (t_k, t_{k+1}]$ $(k = 0, 1, 2, \cdots)$ is considered. Consider the following Lyapunov function for error system (6),

$$V(t) = \int_\Omega \{\frac{1}{2} \sum_{i=1}^{N} {}_{t_0}^R I_t^{1-\alpha}(e_i^T(x,t)e_i(x,t)) + \sum_{i=1}^{N} \frac{1}{1-\tau^D} \int_{t-\tau(t)}^{t} e_i^T(x,s)Qe_i(x,s)ds\}dx, \tag{11}$$

where $Q > 0$ and $Q \in R^{n \times n}$.

Taking the time derivative of $V(t)$ along the trajectories of (6), it is evident that $e_i(x,t)$ and $\frac{\partial^\alpha e_i(x,t)}{\partial t^\alpha}$ are continuous when $(x,t) \in \Omega \times (t_k, t_{k+1}]$. Then, by Lemmas 2 and 3, the following inequality holds:

$$\dot{V}(t) \leq \int_\Omega \{ \sum_{i=1}^N [e_i^T(x,t) \frac{\partial^\alpha e_i(x,t)}{\partial t^\alpha}] + \sum_{i=1}^N [\frac{1}{1-\tau^D} e_i^T(x,t) Qe_i(x,t)$$

$$- \frac{1-\dot{\tau}(t)}{1-\tau^D} e_i^T(x,t-\tau(t)) Qe_i(x,t-\tau(t))] \} dx$$

$$\leq \int_\Omega \{ \sum_{i=1}^N [e_i^T(x,t) D\Delta e_i(x,t) + e_i^T(x,t)(f(z_i(x,t)) - f(z_0(x,t)))$$

$$- ce_i^T(x,t) \sum_{j=1}^N L_{ij}\Gamma e_j(x,t) - \hat{c}e_i^T(x,t) \sum_{j=1}^N \hat{L}_{ij}\hat{\Gamma} e_j(x,t-\tau_j(t)) + u_i(x,t)]$$

$$+ \sum_{i=1}^N [\frac{1}{1-\tau^D} e_i^T(x,t) Qe_i(x,t) - e_i^T(x,t-\tau(t)) Qe_i(x,t-\tau(t))] \} dx \qquad (12)$$

$$= \int_\Omega \sum_{i=1}^N e_i^T(x,t) D\Delta e_i(x,t) dx + \int_\Omega \sum_{i=1}^N e_i^T(x,t)[f(z_i(x,t)) - f(z_0(x,t))] dx$$

$$- \int_\Omega \sum_{i=1}^N [ce_i^T(x,t) \sum_{j=1}^N L_{ij}\Gamma e_j(x,t) + \hat{c}e_i^T(x,t) \sum_{j=1}^N \hat{L}_{ij}\hat{\Gamma} e_j(x,t-\tau(t)) - u_i(x,t)] dx$$

$$+ \int_\Omega \sum_{i=1}^N [\frac{1}{1-\tau^D} e_i^T(x,t) Qe_i(x,t) - e_i^T(x,t-\tau(t)) Qe_i(x,t-\tau(t))] dx.$$

From Green's formula and the boundary conditions, the following equality holds:

$$\int_\Omega \sum_{i=1}^N e_i^T(x,t) D\Delta e_i(x,t) dx = \int_\Omega \sum_{i=1}^N \sum_{l=1}^n e_{il}(x,t) d_l \Delta e_{il}(x,t) dx$$

$$= \sum_{i=1}^N \sum_{l=1}^n d_l \int_\Omega e_{il}(x,t) \Delta e_{il}(x,t) dx \qquad (13)$$

$$= \sum_{i=1}^N \sum_{l=1}^n d_l [-\sum_{k=1}^q \int_\Omega (\frac{\partial e_{il}(x,t)}{\partial x_k})^2] dx.$$

According to Lemma 4, the following inequality can be obtained:

$$\sum_{i=1}^N \sum_{l=1}^n d_l [-\sum_{k=1}^q \int_\Omega (\frac{\partial e_{il}(x,t)}{\partial x_k})^2] dx \leq \sum_{i=1}^N \sum_{l=1}^n d_l [-\sum_{k=1}^q \frac{1}{l_k^2} \int_\Omega e_{il}^2(x,t)] dx$$

$$= -\sum_{i=1}^N \sum_{k=1}^q \frac{1}{l_k^2} \sum_{l=1}^n d_l \int_\Omega e_{il}^2(x,t) dx \qquad (14)$$

$$= -\sum_{i=1}^N \sum_{k=1}^q \frac{1}{l_k^2} \int_\Omega e_i^T(x,t) De_i(x,t) dx.$$

It follows from Assumption 3 and inequalities (12)–(14) that

$$\dot{V}(t) \leq -\sum_{i=1}^N \sum_{k=1}^q \frac{1}{l_k^2} \int_\Omega e_i^T(x,t) De_i(x,t) dx + \sum_{i=1}^N \int_\Omega e_i^T(x,t) \Xi e_i(x,t) dx$$

$$- \int_\Omega [ce^T(x,t)(L \otimes \Gamma)e(x,t) + \hat{c}e^T(x,t)(\hat{L} \otimes \hat{\Gamma})e(x,t-\tau(t))] dx \qquad (15)$$

$$- \int_\Omega \sum_{i=1}^N e_i^T(x,t) b_i \Gamma e_i(x,t) dx + \int_\Omega \sum_{i=1}^N [\frac{1}{1-\tau^D} e_i^T(x,t) Qe_i(x,t) - e_i^T(x,t-\tau(t)) Qe_i(x,t-\tau(t))] dx,$$

where $e(x,t-\tau(t)) = (e_1^T(x,t-\tau(t)), e_2^T(x,t-\tau(t)), \cdots, e_N^T(x,t-\tau(t)))^T$.

By using Lemma 6, an inequality holds as follows:

$$
\begin{aligned}
\dot{V}(t) \leq & -\int_{\Omega} \zeta e^T(x,t)(I_N \otimes D)e(x,t)dx + \int_{\Omega} e^T(x,t)(I_N \otimes \Xi)e(x,t)dx - \int_{\Omega} ce^T(x,t)(L \otimes \Gamma)e(x,t)dx \\
& + \int_{\Omega} \frac{\hat{c}}{2}[\hat{e}e^T(x,t)(\hat{L} \otimes \hat{\Gamma})^2 e(x,t) + \frac{1}{\epsilon}e^T(x,t-\tau(t))e(x,t-\tau(t))]dx - \int_{\Omega} e^T(x,t)(B \otimes \Gamma)e(x,t)dx \\
& + \int_{\Omega}[\frac{1}{1-\tau^D}e^T(x,t)(I_N \otimes Q)e(x,t) - e^T(x,t-\tau(t))(I_N \otimes Q)e(x,t-\tau(t))]dx \\
= & \int_{\Omega} e^T(x,t)[I_N \otimes (-\zeta D + \Xi + \frac{1}{1-\tau^D}Q) - (cL+B) \otimes \Gamma + \frac{\epsilon\hat{c}}{2}(\hat{L} \otimes \hat{\Gamma})^2]e(x,t)dx \\
& + \int_{\Omega} e^T(x,t-\tau(t))[I_N \otimes (\frac{\hat{c}}{2\epsilon}I_n - Q)]e(x,t-\tau(t))dx,
\end{aligned}
\tag{16}
$$

where $\zeta = \sum_{k=1}^{q} \frac{1}{l_k^2}$, $B = diag(b_1, b_2, \cdots, b_N)$ and $b_i \geq 0$ $(i = 1,2,\cdots,l)$, $b_i = 0$ $(i = l+1,\cdots,N)$.

Then, according to Lemma 5 and conditions (8) and (9) in Theorem 1, the following inequality holds:

$$
\dot{V}(t) \leq \int_{\Omega} e^T(x,t)[I_N \otimes (-\zeta D + \Xi + \frac{1}{1-\tau^D}Q) - (cL+B) \otimes \Gamma + \frac{\epsilon\hat{c}}{2}(\hat{L} \otimes \hat{\Gamma})^2]e(x,t)dx < 0. \tag{17}
$$

Next, it can be proven that $\| e(x,t) \| \to 0$ for $t \to \infty$ and $t \neq t_k$. Suppose, for the purpose of contradiction, that $\lim_{t\to\infty} \| e(x,t) \| \neq 0$. According to the properties of function $V(t)$ that $V(t) \geq 0$ and $\dot{V}(t) < 0$, there exists a positive constant $\eta > 0$ such that $V(t) \to \eta$ for $t \to \infty$ (monotone and bounded property). Thus, $\dot{V}(t) \to 0$ for $t \to \infty$. Then the following inequality holds:

$$
0 = \lim_{t\to\infty} \dot{V}(t) \leq \lim_{t\to\infty} \int_{\Omega} e^T(x,t)\Phi e(x,t)dx \leq \lim_{t\to\infty} \int_{\Omega} \lambda_{max}\Phi \| e(x,t) \|^2 dx < 0, \tag{18}
$$

where $\Phi = I_N \otimes (-\zeta D + \Xi + \frac{1}{1-\tau^D}Q) - (cL+B) \otimes \Gamma + \frac{\epsilon\hat{c}}{2}(\hat{L} \otimes \hat{\Gamma})^2$. This contradicts the previous hypothesis.

In conclusion, $\| e(x,t) \| \to 0$ for $t \to \infty$. According to Definition 3, complex network (2) can be synchronized under controller (7). □

Remark 4. *According to Lemma 5, the matrix Q in (8) and (9) can be found by solving the linear matrix inequalities $\lambda_{max}(-\zeta D + \Xi + \frac{1}{1-\tau^D}Q)I_N - \lambda_{min}(\Gamma)(cL+B) + \frac{\epsilon\hat{c}\lambda_{max}^2(\hat{\Gamma})}{2}\hat{L}^2 < 0$ and $\frac{\hat{c}}{2\epsilon}I_n - Q \leq 0$. Furthermore, conditions (8) and (9) provide the control design for synchronization of network (2). It is easy to see that the coupling strengths c and $\hat{c}$ play a key role in (8) and (9). If the coupling strengths c and $\hat{c}$ can be designed, the larger c and the smaller $\hat{c}$ can make these conditions easier to satisfy.*

In order to reduce the enormous difference in control strength between theoretical values and practical need, adaptive and pinning control is considered. Then, without loss of generality, it is assumed that the first l nodes are controlled in the complex networks (2) and an adaptive pinning controller is designed by

$$
\begin{cases}
u_i(x,t) = -b_i(t)\Gamma e_i(x,t) & i = 1,\ldots,l, \\
\dot{b}_i(t) = \int_{\Omega} e_i^T(x,t)\Gamma e_i(x,t)dx,
\end{cases}
\tag{19}
$$

where $b_i(t) \geq 0$ $(i = 1,2,\cdots,l)$ denote control strength.

Next, a theorem is obtained to guarantee that the complex networks (2) can be adaptively synchronized.

Theorem 2. *Suppose Assumptions 1–3 hold. Under the adaptive law and controller* (19), *network* (2) *with initial conditions* (4) *and* (5) *can be synchronized if there exists a positive definite matrix $Q \in R^{n \times n}$ and constants $\epsilon > 0$, $b_i^* \geq 0$ $(i = 1, \cdots, l)$ such that the following conditions are satisfied:*

$$\begin{bmatrix} \lambda_{max}(-\xi D + \Xi + \frac{1}{1-\tau^D}Q)I_N - \lambda_{min}(\Gamma)(cL + B^*) & \frac{\sqrt{2\epsilon \hat{c}}\lambda_{max}(\hat{\Gamma})}{2}\hat{L} \\ \frac{\sqrt{2\epsilon \hat{c}}\lambda_{max}(\hat{\Gamma})}{2}\hat{L} & -I_N \end{bmatrix} < 0, \tag{20}$$

$$\frac{\hat{c}}{2\epsilon}I_n - Q \leq 0, \tag{21}$$

where $B^* = diag(b_1^*, b_2^*, \cdots, b_N^*)$ and $b_i^* \geq 0$ $(i = 1, \cdots, l)$, $b_i^* = 0$ $(i = l+1, \cdots, N)$. $\xi = \sum_{k=1}^{q} \frac{1}{l_k^2}$.

Proof of Theorem 2. The proof of this theorem is given in two steps.

Firstly, the case of $t > t_0$ and $t = t_k$, $(k = 1, 2, \cdots)$ is considered. Similar to the proof of Theorem 1, the following conclusion can be obtained immediately:

$$\| e(x, t_k^+) \|^2 \leq \| e(x, t_k) \|^2. \tag{22}$$

Secondly, the case of $t \geq t_0$ and $t \in (t_k, t_{k+1}]$ $(k = 0, 1, 2, \cdots)$ is considered. Consider the following Lyapunov function for error system (6),

$$V(t) = \int_{\Omega} \{\frac{1}{2}\sum_{i=1}^{N} {}_{t_0}^R I_t^{1-\alpha}(e_i^T(x, t)e_i(x, t)) + \sum_{i=1}^{N} \frac{1}{1-\tau^D}\int_{t-\tau(t)}^{t} e_i^T(x, s)Qe_i(x, s)ds\}dx + \frac{1}{2}\sum_{i=1}^{N}(b_i(t) - b_i^*)^2, \tag{23}$$

where $Q > 0 \in R^{n \times n}$ and $b_i^* \geq 0$ $(i = 1, \cdots, l)$, $b_i^* = 0$ $(i = l+1, \cdots, N)$. b_i^* $(i = 1, \cdots, l)$ are non-negative constants that should be determined later.

Similar to the proof of Theorem 1, computing the derivative of $V(t)$ along the trajectories of error system (6) under the adaptive law and controller (19), the following inequality is obtained,

$$\begin{aligned}
\dot{V}(t) \leq &-\sum_{i=1}^{N}\sum_{k=1}^{q}\frac{1}{l_k^2}\int_{\Omega}e_i^T(x,t)De_i(x,t)dx + \sum_{i=1}^{N}\int_{\Omega}e_i^T(x,t)\Xi e_i(x,t)dx \\
&-\int_{\Omega}[ce^T(x,t)(L \otimes \Gamma)e(x,t) + \hat{c}e^T(x,t)(\hat{L} \otimes \hat{\Gamma})e(x,t-\tau(t))]dx \\
&-\int_{\Omega}\sum_{i=1}^{N}e_i^T(x,t)b_i(t)\Gamma e_i(x,t)dx + \int_{\Omega}\sum_{i=1}^{N}[\frac{1}{1-\tau^D}e_i^T(x,t)Qe_i(x,t) \\
&-e_i^T(x,t-\tau(t))Qe_i(x,t-\tau(t))]dx + \sum_{i=1}^{N}(b_i(t)-b_i^*)\int_{\Omega}e_i^T(x,t)\Gamma e_i(x,t)dx.
\end{aligned} \tag{24}$$

Then,

$$\begin{aligned}
\dot{V}(t) = &-\sum_{i=1}^{N}\sum_{k=1}^{q}\frac{1}{l_k^2}\int_{\Omega}e_i^T(x,t)De_i(x,t)dx + \sum_{i=1}^{N}\int_{\Omega}e_i^T(x,t)\Xi e_i(x,t)dx \\
&-\int_{\Omega}[ce^T(x,t)(L \otimes \Gamma)e(x,t) + \hat{c}e^T(x,t)(\hat{L} \otimes \hat{\Gamma})e(x,t-\tau(t))]dx \\
&-\int_{\Omega}\sum_{i=1}^{N}e_i^T(x,t)b_i^*\Gamma e_i(x,t)dx + \int_{\Omega}\sum_{i=1}^{N}[\frac{1}{1-\tau^D}e_i^T(x,t)Qe_i(x,t) \\
&-e_i^T(x,t-\tau(t))Qe_i(x,t-\tau(t))]dx.
\end{aligned} \tag{25}$$

The rest is the same as in the proof of Theorem 1. Therefore, the complex networks (2) can be synchronized under the adaptive law and controller (19). Meanwhile, the following equalities hold:

$$\lim_{t\to\infty} b_i(t) = b_i^*, \quad for \ \lim_{t\to\infty} \dot{b}_i(t) = 0, \ i = 1, \cdots, l. \tag{26}$$

□

3.2. Pinning Scheme of Complex Networks

In order to design a pinning scheme for network (2), some notations are introduced for simplicity. Let

$$\Psi = \lambda_{max}(-\xi D + \Xi + \frac{1}{1-\tau^D}Q)I_N + \frac{\epsilon\hat{c}\lambda_{max}^2(\hat{\Gamma})}{2}\hat{L}^2 - c\lambda_{min}(\Gamma)L. \tag{27}$$

Using matrix decomposition, the following equation holds:

$$H = \Psi - \lambda_{min}(\Gamma)B = \begin{bmatrix} A - \lambda_{min}(\Gamma)\tilde{B} & E \\ E^T & C \end{bmatrix}, \tag{28}$$

where $\tilde{B} = diag(b_1, b_2, \cdots, b_l)$. $C = (\lambda_{max}(-\xi D + \Xi + \frac{1}{1-\tau^D}Q)I_N + \frac{\epsilon\hat{c}\lambda_{max}^2(\hat{\Gamma})}{2}\hat{L}^2 - c\lambda_{min}(\Gamma)L)_l$ is obtained by removing the first l row–column pairs of matrix Ψ.

Then, a necessary condition is proposed to clearly reveal how the network's characters can affect the pinning synchronization criteria.

Theorem 3. *(Necessary condition) Suppose Assumptions 1–3 hold. To satisfy condition (8), it is necessary that*

$$H_{ii} = \lambda_{max}(-\xi D + \Xi + \frac{1}{1-\tau^D}Q) + \frac{\epsilon\hat{c}\lambda_{max}^2(\hat{\Gamma})}{2}\sum_{j=1}^{N}\hat{L}_{ij}^2 - \lambda_{min}(\Gamma)(cL_{ii} + b_i) < 0, \ 1 \le i \le l, \tag{29}$$

$$H_{ii} = \lambda_{max}(-\xi D + \Xi + \frac{1}{1-\tau^D}Q) + \frac{\epsilon\hat{c}\lambda_{max}^2(\hat{\Gamma})}{2}\sum_{j=1}^{N}\hat{L}_{ij}^2 - \lambda_{min}(\Gamma)cL_{ii} < 0, \ l+1 \le i \le N. \tag{30}$$

Proof of Theorem 3. According to Lemma 5, condition (8) is equivalent to $H < 0$. It is necessary that $H_{ii} < 0$. □

Remark 5. $\hat{L}_{ij}$ *is an element of the Laplacian matrix $\hat{L}$, which denotes the delayed communication topology of network (2). According to the definition of the Laplacian matrix, the greater the degree of the ith node is, the greater the value of $\sum_{j=1}^{N}\hat{L}_{ij}^2$ is, and vice versa. From (29) and (30), for the nodes without a controller, the degrees of these nodes in the delayed communication topology must be less than a critical value. Namely, the condition $\sum_{j=1}^{N}\hat{L}_{ij}^2 < (\frac{2}{\epsilon\hat{c}\lambda_{max}^2(\hat{\Gamma})})(\lambda_{min}(\Gamma)cL_{ii} - \lambda_{max}(-\xi D + \Xi + \frac{1}{1-\tau^D}Q))$ must be satisfied, which indicates that the nodes with large degrees in delayed communication topology should be controlled first, otherwise, (30) is not satisfied. This is consistent with the intuition that the delayed communication between nodes can lead to some instability in the network. In addition, some results have been proposed [43] that, in the instant communication topology network, nodes with very low and very large degrees are good candidates for applying pinning controllers.*

In order to be more practical, some low-dimensional conditions are presented to guarantee the global asymptotic stability of the pinning process.

Theorem 4. *Suppose Assumptions 1–3 hold. Under controller* (7), *the pinning controlled network* (2) *is globally synchronized when the following two conditions are satisfied:*

$$b_i > \frac{\lambda_{max}(A - EC^{-1}E^T)}{\lambda_{min}(\Gamma)}, \quad i = 1, 2, \cdots, l, \tag{31}$$

$$\lambda_{max}\left(\frac{\epsilon \hat{c} \lambda_{max}^2(\hat{\Gamma})}{2} \hat{L}^2 - c\lambda_{min}(\Gamma)L\right)_l < -\lambda_{max}(-\xi D + \Xi + \frac{1}{1 - \tau^D}Q), \tag{32}$$

where b_i is the pinning feedback gain, A, B, and C are defined in (28), *and* $\left(\frac{\epsilon \hat{c} \lambda_{max}^2(\hat{\Gamma})}{2} \hat{L}^2 - c\lambda_{min}(\Gamma)L\right)_l$ *is the minor matrix of* $\frac{\epsilon \hat{c} \lambda_{max}^2(\hat{\Gamma})}{2} \hat{L}^2 - c\lambda_{min}(\Gamma)L$ *by removing its first l row–column pairs.*

Proof of Theorem 4. From Lemma 5, it follows (31) and (32) that $H < 0$, which is equivalent to (8). □

Remark 6. *The synchronization conditions* (31) *and* (32) *are easier to verify than* (8). *Thus, Theorem 4 has more practical value. Meanwhile, the conditions* (8), (31), *and* (32) *provide a method of controller design for synchronization of the complex networks* (2).

Remark 7. *Note that for $\alpha = 1$, the model of complex networks* (2) *reduces to the classical integer-order system with impulsive effects and reaction–diffusion terms. It is not difficult to verify that, in the case of $\alpha = 1$, the conditions given in Theorems 1, 2, and 4 can also guarantee the synchronization of corresponding integer-order complex networks via the designed controllers* (7) *and* (19). *Therefore, the synchronization results in Theorems 1, 2, and 4 extend and improve the synchronization results for integer-order complex networks with reaction–diffusion terms and impulsive effects compared to the fractional case.*

Remark 8. *When $D = 0$, the partial differential equations* (2) *can be degraded to fractional ordinary differential equations. It is easy to demonstrate that in this special case, the criteria given in Theorems 1, 2, and 4 are also valid for ensuring the synchronization of corresponding ordinary differential systems under controllers* (7) *and* (19).

Remark 9. *In recent years, many outstanding results from the analysis of the stability of fractional systems, as in* [46–49], *have been reported. But, since there exists an integration term in the definition of the Riemann–Liouville fractional derivative, which means that the fractional derivative of a function $x(t)$ at any given moment depends on its initial state, the existing fractional Lyapunov methods are invalid for analyzing the stability of fractional systems with impulsive effects. Thus, according to Lemma 2, two special Lyapunov functions with fractional integration terms are constructed to complete the proof of Theorems 1 and 2.*

Remark 10. *Adaptive control for fractional complex networks* (2) *is studied in this paper. However, there is no such property of the Riemann–Liouville fractional derivative that $x(t) \to m$ for $_{t_0}^R D_t^\alpha x(t) \to 0$, where m is a constant. Therefore, an integer-order adaptive law* (19) *is designed rather than a fractional-order adaptive law. This is more practical because the computational complexity of fractional derivatives is much higher than that of integral derivatives.*

4. Numerical Simulations

In this section, an example is given to illustrate the effectiveness of the main results.

Example 1. *Consider the two-dimensional Riemann–Liouville fractional complex network* (2), *which consists of five nodes, with impulses and reaction–diffusion terms. The parameters are designed as: $\alpha = 0.95$; $\Delta = \frac{\partial^2}{\partial x}$, $x \in \Omega = \{x \mid 0 \le x \le \pi\}$; $D = diag(0.3, 0.3)$ $\tau(t) = 1$; $\Gamma, \hat{\Gamma}$ are identity matrices; $c = 1$, $\hat{c} = 0.25$; and the Laplacian matrices $L, \hat{L}$ are given as follows:*

$$
L = \begin{bmatrix} 0.1 & -0.1 & 0 & 0 & 0 \\ -0.1 & 0.2 & -0.1 & 0 & 0 \\ 0 & -0.1 & 0.2 & -0.1 & 0 \\ 0 & 0 & -0.1 & 0.2 & -0.1 \\ 0 & 0 & 0 & -0.1 & 0.1 \end{bmatrix}, \quad \hat{L} = \begin{bmatrix} 0.2 & -0.2 & 0 & 0 & 0 \\ -0.2 & 0.4 & -0.2 & 0 & 0 \\ 0 & -0.2 & 0.4 & -0.2 & 0 \\ 0 & 0 & -0.2 & 0.4 & -0.2 \\ 0 & 0 & 0 & -0.2 & 0.2 \end{bmatrix}. \tag{33}
$$

The nonlinear function $f : R^2 \rightarrow R^2$ is

$$
f(s(t)) = \begin{pmatrix} 2tanh(s_1(t)) - 1.2tanh(s_2(t)) \\ 1.8tanh(s_1(t)) + 1.71tanh(s_2(t)) \end{pmatrix},
$$

where $s(t) = (s_1(t), s_2(t))^T$. The boundary and initial value conditions of the isolated node (3) and network (2) are given in the form

$$
z_i(x, t) = 0, \quad t \in [-1, \infty), \quad x \in \partial\Omega, \quad i = 0, 1, 2, \cdots, 5; \tag{34}
$$

$$
z_i(x, s) = (\varphi_{i1}(x), \varphi_{i2}(x))^T, \quad s \in [-1, 0), \quad x \in \Omega, \quad i = 0, 1, 2, \cdots, 5, \tag{35}
$$

where $\varphi_{01}(x) = -sin(x)$, $\varphi_{02}(x) = 0.3sin(x)$, $\varphi_{11}(x) = \varphi_{12}(x) = sinx$, $\varphi_{21}(x) = \varphi_{22}(x) = 0.5sinx$, $\varphi_{31}(x) = \varphi_{32}(x) = -0.7sinx$, $\varphi_{41}(x) = \varphi_{42}(x) = -0.5sinx$, $\varphi_{51}(x) = \varphi_{52}(x) = 1.5sinx$. The impulsive strength functions are given as follows: $\delta_{1k}(x, t_k) = 0.8$, $\delta_{2k}(x, t_k) = 0.8 \mid cos(\frac{x}{4})sin(\frac{t_k}{4}) \mid$, $\delta_{3k}(x, t_k) = 0.3 \mid cos(xt_k) \mid$, $\delta_{4k}(x, t_k) = 0.1 \mid sin(\frac{x}{2})cos(\frac{t_k}{5}) \mid$, $\delta_{5k}(x, t_k) = 0.2 \mid sin(x)cos(t_k) \mid$, $k = 1, 2, \cdots$. The pulse period is $t_{k+1} - t_k = 0.4s$.

Pinning control strategy is considered here, supposing the first node in network (2) is controlled. Then, the appropriate controller and adaptive law can be designed as follows:

$$
\begin{cases} u_i(x, t) = -b_i(t)\Gamma e_i(x, t) & i = 1, \\ \dot{b}_i(t) = \int_\Omega e_i^T(x, t)\Gamma e_i(x, t)dx, \end{cases} \tag{36}
$$

where $e_i(x, t) = z_i(x, t) - z_0(x, t)$ $(i = 1, \cdots, 5)$. $b_1(t) > 0$ and $b_1(0) = 0.2$.

Let $\parallel e(x, t) \parallel$ stand for the norm of synchronization error between systems (2) and (3). In Figures 1–4, it is shown that under the adaptive law and designed controller (36), complex networks (2) with reaction–diffusion terms and impulsive effects can achieve synchronization. In Figure 5, it is easy to see that the adaptive control parameter $b_1(t)$ turns out to be a constant $b_1^* = 8.788$, which satisfies the conditions in Theorem 2, when synchronization is realized. The efficiency of Theorem 2 can be demonstrated by this example.

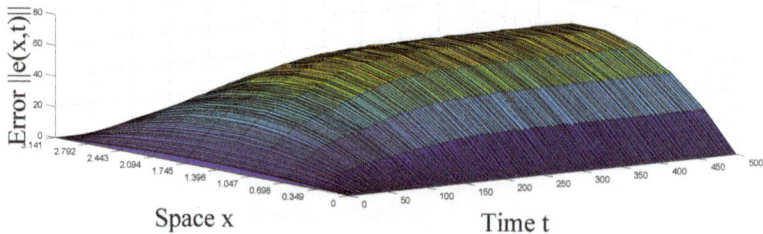

Figure 1. Spatiotemporal evolution of the error norm $\parallel e(x, t) \parallel$ between systems (2) and (3) without controller.

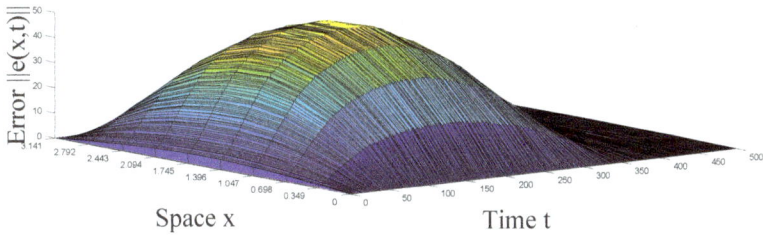

Figure 2. Spatiotemporal evolution of the error norm $\| e(x,t) \|$ between systems (2) and (3) with controller (36).

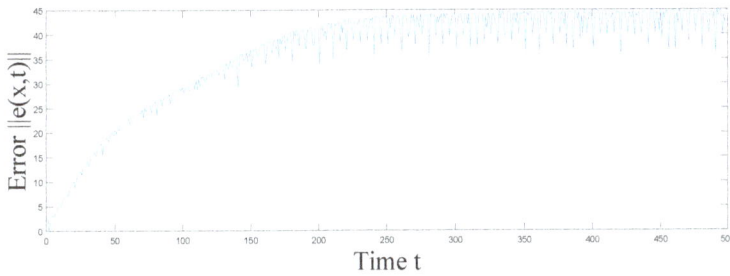

Figure 3. Time evolution of the synchronization error $\| e(x,t) \|$ between systems (2) and (3) without controller at $x = 0.698$.

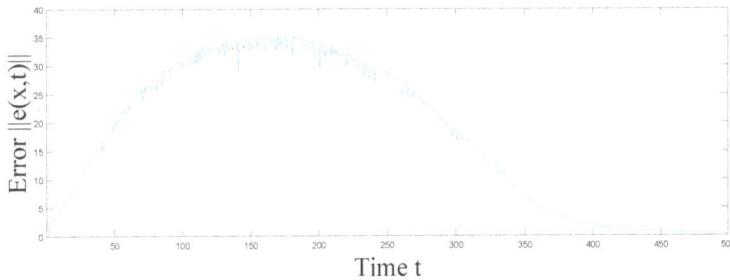

Figure 4. Time evolution of the synchronization error $\| e(x,t) \|$ between systems (2) and (3) with controller (36) at $x = 0.698$.

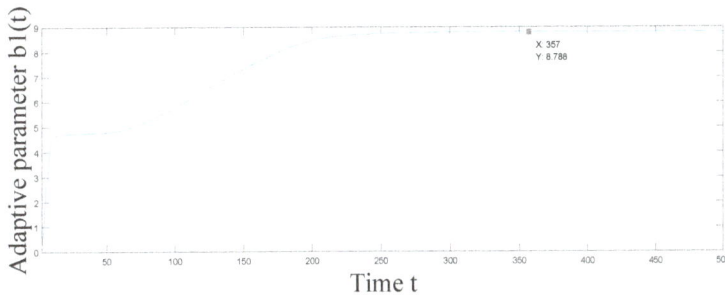

Figure 5. Trajectory of control parameter $b_1(t)$ of the adaptive controller (36) in Example 1.

5. Conclusions

Pinning and adaptive synchronization of fractional complex networks with impulses and reaction–diffusion terms is investigated in this paper. In order to analyze the stability of the fractional systems with impulsive effects, two special Lyapunov functions with fractional integration terms are constructed and the Lyapunov method is applied. By designing appropriate controllers and adaptive laws, some sufficient criteria for pinning and adaptive synchronization of fractional complex networks with time-varying delays are derived. In addition, a pinning scheme is proposed in which some delay-coupled nodes should be prioritized for pinning. Finally, a numerical example is given to demonstrate the effectiveness and correctness of the main results. A goal of our future investigations is to study more general impulsive fractional systems and analyze the dynamical behaviors of these special systems.

Author Contributions: Writing—original draft preparation, X.H.; writing—review and editing, G.R., Y.Y. and C.X.

Funding: This research is supported by the National Natural Science Foundation of China (No. 61772063) and Beijing Natural Science Foundation (Z180005).

Conflicts of Interest: The authors declare no conflict of interest.

References

1. Bernardo, A.H.; Lada, A.A. Internet: Growth dynamics of the World-Wide Web. *Nature* **1999**, *401*, 131.
2. Ding, W.Y.; Wei, X.C.; Yi, D. A closed-form solution for the impedance calculation of grid power distribution network. *IEEE Trans. Electromagn. Compat.* **2017**, *59*, 1449–1456. [CrossRef]
3. Strogatz, S.H. Exploring complex networks. *Nature* **2001**, *410*, 268–276. [CrossRef]
4. Selvaraj, P.; Sakthivel, R.; Kwon, O.M. Synchronization of fractional-order complex dynamical network with random coupling delay, actuator faults and saturation. *Nonlinear Dyn.* **2018**, *94*, 3101–3116. [CrossRef]
5. Yih-Fang, H.; Stefan, W.; Jing, H.; Neelabh, K.; Vijay, G. State estimation in electric power grids: Meeting new challenges presented by the requirements of the future grid. *IEEE Signal Process. Mag.* **2012**, *29*, 33–43.
6. Luo, S.X.; Deng, F.Q.; Chen, W.H. Pointwise-in-space stabilization and synchronization of a class of reaction-diffusion systems with mixed time delays via aperiodically impulsive control. *Nonlinear Dyn.* **2017**, *88*, 2899–2914. [CrossRef]
7. Mo, L.P.; Lin, P. Distributed consensus of second-order multiagent systems with nonconvex input constraints. *Int. J. Robust Nonlinear Control.* **2018**, *28*, 3657–3664. [CrossRef]
8. Mo, L.P.; Guo, S.Y.; Yu, Y.G. Mean-square consensus of heterogeneous multi-agent systems with nonconvex constraints, Markovian switching topologies and delays. *Neurocomputing* **2018**, *291*, 167–174. [CrossRef]
9. Wang, K.H.; Fu, X.C.; Li, K.Z. Cluster synchronization in community networks with nonidentical nodes. *Chaos* **2009**, *19*, 023106. [CrossRef]
10. Liu, S.T.; Zhang, F.F. Complex function projective synchronization of complex chaotic system and its applications in secure communication. *Nonlinear Dyn.* **2014**, *76*, 1087–1097. [CrossRef]
11. Li, C.G.; Chen, G.R. Synchronization in general complex dynamical networks with coupling delays. *Physica A* **2004**, *343*, 263–278. [CrossRef]
12. Wu, X.J.; Lu, H.T. Projective lag synchronization of the general complex dynamical networks with distinct nodes. *Commun. Nonlinear Sci. Numer. Simul.* **2012**, *17*, 4417–4429. [CrossRef]
13. Yu, C.B.; Qin, J.H.; Gao, H.J. Cluster synchronization in directed networks of partial-state coupled linear systems under pinning control. *Automatica* **2014**, *50*, 2341–2349. [CrossRef]
14. Yi, J.W.; Wang, Y.W.; Xiao, J.W.; Huang, Y.H. Exponential synchronization of complex dynamical networks with Markovian jump parameters and stochastic delays and its application to multi-agent systems. *Commun. Nonlinear Sci. Numer. Simul.* **2013**, *18*, 1175–1192. [CrossRef]
15. Selvaraj, P.; Kwon, O.M.; Sakthivel, R. Disturbance and uncertainty rejection performance for fractional-order complex dynamical networks. *Neural Netw.* **2019**, *112*, 73–84. [CrossRef]
16. Chen, J.; Lu, J.A.; Wu, X.Q.; Zheng, W.X. Generalized synchronization of complex dynamical networks via impulsive control. *Chaos* **2009**, *19*, 043119. [CrossRef]

17. Zhang, Q.; Lu, J.A.; Lu, J.H.; Tse, C.K. Adaptive feedback synchronization of a general complex dynamical network with delayed nodes. *IEEE Trans. Circuits Syst. II Express Briefs* **2008**, *55*, 183–187. [CrossRef]
18. Chen, H.W.; Liang, J.L.; Wang, Z.D. Pinning controllability of autonomous Boolean control networks. *Sci. China-Inf. Sci.* **2016**, *59*, 070107. [CrossRef]
19. Liu, X.W.; Chen, T.P. Synchronization of complex networks via aperiodically intermittent pinning control. *IEEE Trans. Autom. Control.* **2015**, *60*, 3316–3321. [CrossRef]
20. Wang, G.S.; Xiao, J.W.; Wang, Y.W.; Yi, J.W. Adaptive pinning cluster synchronization of fractional-order complex dynamical networks. *Appl. Math. Comput.* **2014**, *231*, 347–356. [CrossRef]
21. Chai, Y.; Chen, L.P.; Wu, R.C.; Sun, J. Adaptive pinning synchronization in fractional-order complex dynamical networks. *Physica A* **2012**, *391*, 5746–5758. [CrossRef]
22. Wang, J.L.; Wu, H.N.; Guo, L. Novel adaptive strategies for synchronization of linearly coupled neural networks with reaction-diffusion terms. *IEEE Trans. Neural Netw. Learn. Syst.* **2014**, *25*, 429–440. [CrossRef]
23. Wang, J.L.; Wu, H.N. Passivity of delayed reaction-diffusion networks with application to a food Web model. *Appl. Math. Comput.* **2013**, *219*, 11311–11326. [CrossRef]
24. Wang, J.L.; Wu, H.N.; Huang, T.W.; Ren, S.Y.; Wu, J.G. Pinning controlfor synchronization of coupled reaction-diffusion neural networks with directed topologies. *IEEE Trans. Syst. Man Cybern. Syst.* **2016**, *46*, 1109–1120. [CrossRef]
25. Song, Q.K.; Yan, H.; Zhao, Z.J.; Liu, Y.R. Global exponential stability of complex-valued neural networks with both time-varying delays and impulsive effects. *Neural Netw.* **2016**, *79*, 108–116. [CrossRef]
26. Zhu, Q.X.; Cao, J.D. Stability analysis of Markovian jump stochastic BAM neural networks with impulsive control and mixed time delays. *Nature* **2012**, *13*, 2259–2270.
27. Keith, B.O.; Jerome, S. *The Fractional Calculus*; Academic Press: New York, NY, USA, 1974.
28. Ionescu, C.; Lopes, A.; Copot, D.; Machado, J.A.T.; Bates, J.H.T. The role of fractional calculus in modeling biological phenomena: A review. *Commun. Nonlinear Sci. Numer. Simul.* **2017**, *51*, 141–159. [CrossRef]
29. Magin, R.; Ortigueira, M.D.; Podlubny, I.; Trujillo, J. On the fractional signals and systems. *Signal Process.* **2011**, *91*, 350–371. [CrossRef]
30. Park, M.; Lee, S.H.; Kwon, O.M.; Seuret, A. Closeness-Centrality-Based Synchronization Criteria for Complex Dynamical Networks With Interval Time-Varying Coupling Delays. *IEEE Trans. Cybern.* **2018**, *48*, 2192–2202. [CrossRef]
31. Li, X.D.; Zhang, X.L.; Song, S.J. Effect of delayed impulses on input-to-state stability of nonlinear systems. *Automatica* **2017**, *76*, 378–382. [CrossRef]
32. Wang, H.; Yu, Y.G.; Wen, G.G.; Zhang, S.; Yu, J.Z. Global stability analysis of fractional-order Hopfield neural networks with time delay. *Neurocomputing* **2015**, *154*, 15–23. [CrossRef]
33. Kilbas, A.A.; Srivastava, H.M.; Trujillo, J.J. *Theory and Applications of Fractional Differential Equations*; Elsevier: Amsterdam, The Netherlands, 2006.
34. Stamova, I. Global Mittag-Leffler stability and synchronization of impulsive fractional-order neural networks with time-varying delays. *Nonlinear Dyn.* **2014**, *77*, 1251–1260. [CrossRef]
35. Lu, J.G. Global exponential stability and periodicity of reaction-diffusion delayed recurrent neural networks with Dirichlet boundary conditions. *Chaos* **2008**, *35*, 116–125. [CrossRef]
36. Boyd, S.; El Ghaoui, L.; Feron, E.; Balakrishnan, V. Linear Matrix Inequalities in System and Control Theory. In *Semidefinite Programming and Linear Matrix Inequalities*; Society for Industrial and Applied Mathematics: Philadelphia, PA, USA, 1994.
37. Liu, S.; Li, X.Y.; Zhou, X.F.; Jiang, W. Synchronization analysis of singular dynamical networks with unbounded time-delays. *Adv. Differ. Equations* **2015**, *2015*, 193. [CrossRef]
38. Chen, T.P.; Liu, X.W.; Lu, W.L. Pinning complex networks by a single controller. *IEEE Trans. Circuits Syst. I Regul. Pap.* **2007**, *54*, 1317–1326. [CrossRef]
39. Roger, A.H.; Charles, R.J. *Topics in Matrix Analysis*; Cambridge University Press: Cambridge, UK, 1994.
40. Roger, A.H.; Charles, R.J. *Matrix Analysis*; Cambridge University Press: Cambridge, UK, 1985.
41. Zhang, W.B.; Wang, Z.D.; Liu, Y.R.; Ding, D.R.; Alsaadi, F.E. Event-based state estimation for a class of complex networks with time-varying delays: A comparison principle approach. *Phys. Lett. A* **2017**, *381*, 10–18. [CrossRef]

42. Feng, J.W.; Yang, P.; Zhao, Y. Cluster synchronization for nonlinearly time-varying delayed coupling complex networks with stochastic perturbation via periodically intermittent pinning control. *Appl. Math. Comput.* **2016**, *291*, 52–68. [CrossRef]

43. Lu, R.Q.; Yu, W.W.; Lu, J.H.; Xue, A.K. Synchronization on Complex Networks of Networks. *IEEE Trans. Neural Netw. Learn. Syst.* **2014**, *25*, 2110–2118. [CrossRef]

44. Abbaszadeh, M.; Marquez, H.J. Nonlinear Observer Design for One-Sided Lipschitz Systems. In Proceedings of the 2010 American Control Conference, Baltimore, MD, USA, 30 June–2 July 2010; pp. 5284–5289.

45. Chen, Y.; Yu, W.W.; Tan, S.L.; Zhu, H.H. Synchronizing nonlinear complex networks via switching disconnected topology. *Automatica* **2016**, *70*, 189–194. [CrossRef]

46. Li, Y.; Chen, Y.Q.; Podlubny, I. Mittag-Leffler stability of fractional order nonlinear dynamic systems. *Automatica* **2009**, *45*, 1965–1969. [CrossRef]

47. Li, Y.; Chen, Y.Q.; Podlubny, I. Stability of fractional-order nonlinear dynamic systems: Lyapunov direct method and generalized Mittag-Leffler stability. *Comput. Math. Appl.* **2010**, *59*, 1810–1821. [CrossRef]

48. Rivero, M.; Rogosin, S.V.; Tenreiro Machado, J.A.; Trujillo, J.J. Stability of Fractional Order Systems. *Math. Probl. Eng.* **2013**, *2013*, 356215. [CrossRef]

49. Sabatier, J.; Farges, C.; Trigeassou, J.C. A stability test for non-commensurate fractional order systems. *Syst. Control. Lett.* **2013**, *62*, 739–746. [CrossRef]

mathematics

MDPI

Article

Optimal Randomness in Swarm-Based Search

Jiamin Wei [1,2], YangQuan Chen [2,*], Yongguang Yu [1] and Yuquan Chen [2,3]

[1]   Department of Mathematics, Beijing Jiaotong University, Beijing 100044, China
[2]   School of Engineering, University of California, Merced, 5200 Lake Road, Merced, CA 95343, USA
[3]   Department of Automation, University of Science and Technology of China, Hefei 230027, China
*   Correspondence: ychen53@ucmerced.edu

Received: 23 July 2019; Accepted: 3 September 2019; Published: 6 September 2019

Abstract: Lévy flights is a random walk where the step-lengths have a probability distribution that is heavy-tailed. It has been shown that Lévy flights can maximize the efficiency of resource searching in uncertain environments and also the movements of many foragers and wandering animals have been shown to follow a Lévy distribution. The reason mainly comes from the fact that the Lévy distribution has an infinite second moment and hence is more likely to generate an offspring that is farther away from its parent. However, the investigation into the efficiency of other different heavy-tailed probability distributions in swarm-based searches is still insufficient up to now. For swarm-based search algorithms, randomness plays a significant role in both exploration and exploitation or diversification and intensification. Therefore, it is necessary to discuss the optimal randomness in swarm-based search algorithms. In this study, cuckoo search (CS) is taken as a representative method of swarm-based optimization algorithms, and the results can be generalized to other swarm-based search algorithms. In this paper, four different types of commonly used heavy-tailed distributions, including Mittag-Leffler distribution, Pareto distribution, Cauchy distribution, and Weibull distribution, are considered to enhance the searching ability of CS. Then four novel CS algorithms are proposed and experiments are carried out on 20 benchmark functions to compare their searching performances. Finally, the proposed methods are used to system identification to demonstrate the effectiveness.

Keywords: optimal randomness; swarm-based search; cuckoo search; heavy-tailed distribution; global optimization

1. Introduction

Swarm-based search algorithms have attracted great interest of researchers in fields of computational intelligence, artificial intelligence, optimization, data mining, and machine learning during the last two decades [1]. Moreover, the swarm intelligence algorithms and artificial intelligence have been successfully used in complex real-life applications, such as wind farm decision system, social aware cognitive radio handovers, feature selection, truck scheduling and so on [2–5]. Up to now, a lot of swarm-based search algorithms have been presented, including artificial bee colony (ABC) [6], cuckoo search (CS) [7], firefly algorithm (FA) [8], particle swarm optimization (PSO) [9] and so on.

Among the existing swarm-based search algorithms, CS is presented in terms of the obligate brood parasitic behavior of some cuckoo species and the Lévy flight behavior of some birds and fruit flies. CS searches for new solutions by performing a global explorative random walk together with a local exploitative random walk. One advantage of CS is that its global search utilizes Lévy flights or process, instead of standard random walks. Lévy flights play a critical role in enhancing randomness, as Lévy flights is a random walk where the step-lengths have a probability distribution that is heavy-tailed. At each iteration process, CS firstly searches for new solutions in Lévy flights random walk. Secondly, CS proceeds to obtain new solutions in local exploitative random walk.

After each random walk, a greedy strategy is used to select a better solution from the current and newly generated solutions according to their fitness values. Due to the salient features such as simple concept, limited parameters, and implementation simplicity, CS has aroused extensive attention and has been accepted as a simple but efficient optimization technique for solving optimization problems. Accordingly, many new CS variants have been continuously presented recently [10–14]. However, there is still a lot of space in designing newly improved or enhanced techniques to help to increase the accuracy and convergence speed and enhance the searching stability for the original CS algorithm.

In nature, the movements of many foragers and wandering animals have been shown to follow a Lévy distribution [15] rather than a Gaussian distribution. It is found that foragers frequently take a large step to enhance their searching efficiency since it is the product of natural evolution over millions of years. Inspired by the mentioned natural phenomena, CS is proposed in combination with Lévy, where the step-length is drawn from a heavy-tailed probability distribution and large steps frequently take place flights. In fact, before CS, the idea of Lévy flights was applied in Reference [16] to solve a problem of non-convex stochastic optimization, due to big jumps of the Lévy flights process. In this way, it can enhance the searching ability compared with the Gaussian distribution where large steps seldom happen. More exactly, we have to say the foragers should move following a heavy-tailed distribution since the Lévy distribution is a simple heavy-tailed distribution which is easy to analyze. There are many other heavy-tailed distributions such as the Mittag-Leffler distribution, Pareto distribution, Cauchy distribution and the Weibull distribution, and large steps still frequently happen when using them to generate the steps. For swarm-based optimization algorithms, randomness plays a significant role in both exploration and exploitation or diversification and intensification [17]. Therefore, it is necessary to discuss the optimal randomness in swarm-based search algorithms.

In this paper, we mainly focus on the discussion on the impact of different heavy-tailed distributions on the performance of swarm-based search algorithms. In the study, CS is taken as a representative method of swarm-based optimization algorithms, and the results can be generalized to other swarm-based search algorithms. At first, some basic definitions of the heavy-tailed distributions and how to generate the random numbers according to the given distribution are provided. Then by replacing the Lévy flight with steps generated from other heavy-tailed distributions, four different randomness-enhanced CS algorithms (namely CSML, CSP, CSC, and CSW) are presented by applying Mittag-Leffler distribution, Pareto distribution, Cauchy distribution and Weibull distribution. Finally, dedicated experimental studies are carried out on a test suite of 20 benchmark problems with unimodal, multimodal, rotated and shifted properties to compare the performance of different variant algorithms. The experimental results demonstrate that the four proposed randomness-enhanced CS algorithms show a significant improvement over the original CS algorithm. This suggests that the performance of CS can be improved by means of integrating different heavy-tailed probability distributions rather than Lévy flights into it. At last, an application problem of parameter identification of unknown fractional-order chaotic systems is further considered. Based on the observations and results analysis, the randomness-enhanced CS algorithms are able to exactly identify the unknown specific parameters of the fractional-order system with better effectiveness and robustness. The randomness-enhanced CS algorithms can be regarded as an efficient and promising tool for solving the real-world complex optimization problems besides the benchmark problems.

The remainder of this paper is organized as follows. The principle of the original CS algorithm is described in Section 2. Section 3 gives details of four randomness-enhanced CS algorithms after a brief review of several commonly used heavy-tailed distributions. Experimental results and discussions of randomness-enhanced CS algorithms are presented in Section 4. Finally, Section 5 summarizes the conclusions and future work.

2. Cuckoo Search Algorithm

Cuckoo search (CS), developed by Yang and Deb, is considered to be a simple but promising stochastic nature-inspired swarm-based search algorithm [7,18]. CS is inspired by the intriguing brood parasitism behaviors of some species of cuckoos, and is enhanced by Lévy flights instead of simple isotropic random walks. Cuckoos are considered to be fascinating birds not only for their beautiful sounds but also for their aggressive reproduction strategy. Some cuckoo species lay their eggs in host nests, and at the same time, they may remove host birds' eggs in order to increase the hatching probability of their own eggs. For simplicity in describing the standard CS, there are three idealized rules as follows [7]: (1) Only one egg is laid by each cuckoo bird at a time, and dumped in a randomly chosen nest; (2) The next generations of cuckoos search for new solutions using the best nests with high-quality; (3) The number of available host nests is fixed, and the egg laid by a cuckoo is discovered by the host bird with a probability $P_a \in [0, 1]$. In this condition, the host bird can either remove the egg or simply abandon the nest and build a completely new nest.

The purpose of CS is to substitute a not-so-good solution in the nests with the new and potentially better solutions (cuckoos). At each iteration process, CS employs a balanced combination of a local random walk and the global explorative random walk under control of a switching parameter P_a. A greedy strategy is used after each random walk to select better solutions from the current and newly generated solutions based on their fitness values.

2.1. Lévy Flights Random Walk

At generation t, a global explorative random walk carried out by using Lévy flights can be defined as follows:

$$U_i^t = X_i^t + \alpha \otimes \text{Lévy} \otimes (X_i^t - X_{best}), \tag{1}$$

where U_i^t denotes a new solution generated in Lévy flights random walk, and X_{best} is the best solution obtained so far. $\alpha > 0$ is the step size related to the scales of the problem of interest, X_{best} is the best solution obtained so far, the product $\otimes$ represents entrywise multiplications, and Lévy(λ) is defined according to a simple power-law formula as follows:

$$\text{Lévy}(\lambda) \sim t^{-1-\lambda}, \tag{2}$$

where t is a random variable, $0 < \lambda \leq 2$ is a stability index. Moreover, it is worth mentioning that the well-known Gaussian and Cauchy distribution are its special cases when its stability index λ is respectively set to 2 and 1.

In practice, Lévy(λ) can be updated as follows:

$$\text{Lévy}(\lambda) \sim \frac{\mu}{|v|^{1/\lambda}} \phi, \tag{3}$$

where λ is suggested as 1.5 [18], μ and v are random numbers drawn from a normal distribution with mean of 0 and standard deviation of 1, $\blacksquare(\cdot)$ denotes the gamma function, and ϕ is expressed as:

$$\phi = \left[\frac{\blacksquare(\lambda) \sin\left(\frac{\pi\lambda}{2}\right)}{\blacksquare\left(\frac{1+\lambda}{2}\right) 2^{\frac{\lambda-1}{2}}} \right]^{1/\lambda}. \tag{4}$$

2.2. Local Random Walk

The local random walk can be defined as:

$$U_i^t = X_i^t + r \otimes H(P_a - \epsilon) \otimes (X_j^t - X_k^t), \tag{5}$$

where X_j^t and X_k^t are two different selected random solutions, r and ϵ are two independent random numbers with uniform distribution, and $H(u)$ is a Heaviside function. The local random walk utilizes a far field randomization to generate a substantial fraction of new solutions which are sufficiently far from the current best solution. The pseudo-code of the standard CS algorithm is given in Algorithm 1.

Algorithm 1 Pseudo code of the standard CS algorithm

Input: Population size NP, fraction probability P_a, dimensionality D, the maximum number of
function evaluations Max_FEs, iteration number $t = 1$, objective function $f(X)$.
Output: The best solution.
1: $t = 1$;
2: Generate an initial population of NP host nests X_i^t, $(i = 1, 2, \ldots, NP)$;
3: Evaluate the fitness value of each nest X_i^t;
4: $FES = NP$;
5: Determine the best nest with the best fitness value;
6: **while** FES<Max_FEs **do**

7:    // **Lévy flights random walk**
8:    **for** $i = 1, 2, \ldots, NP$ **do**

9:       Generate a new solution U_i^t randomly using Lévy flights random walk according to
         Equation (1);
10:      Greedily select a better solution from U_i^t and X_i^t according to their fitness values;
11:      $FES = FES + 1$;
12:   **end for**
13:   // **Local random walk, a fraction (P_a) of worse nests are abandoned and new ones are built**
14:   **for** $i = 1, 2, \ldots, NP$ **do**

15:      Search for a new solution U_i^t using local random walk according to Equation (5);
16:      Greedily select a better solution from U_i^t and X_i^t according to their fitness values;
17:      $FES = FES + 1$;
18:   **end for**
19:   Obtain the best solution so far X_{best};
20:   $t = t + 1$;
21: **end while**

---

3. Randomness-Enhanced CS Algorithms

The standard CS algorithm uses Lévy flights in global random walk to explore the search space. The Lévy step is taken from the Lévy distribution which is a heavy-tailed probability distribution. In this case, a fraction of large steps is generated, which plays an important role in enhancing the search capability of CS. Although many foragers and wandering animals have been shown to follow a Lévy distribution [15], the investigation into the impact of other different heavy-tailed probability distributions on CS is still insufficient up to now. This motivates us to make an attempt to apply the well-known Mittag-Leffler distribution, Pareto distribution, Cauchy distribution and Weibull distribution to the standard CS algorithm, by using which, more efficient searches are supposed to take place in the search space thanks to the long jumps. In this section, a brief review of several commonly used heavy-tailed distributions is given and then the scheme of the randomness-enhanced CS algorithms is introduced.

3.1. Commonly Used Heavy-Tailed Distributions

This subsection provides the definition of heavy-tailed distribution and several examples of commonly used heavy-tailed distributions.

Definition 1 (Heavy-Tailed Distribution). *The distribution of a real-valued random variable X is said to have a heavy right tail if the tail probabilities $P(X > x)$ decay more slowly than those of any exponential distribution, that is, if*

$$\lim_{x \to \infty} \frac{P(X > x)}{e^{-\lambda x}} = \infty \tag{6}$$

for every $\lambda > 0$. Heavy left tails are defined in a similar way [19].

Example 1 (Mittag-Leffler Distribution). *A random variable is said to subject to Mittag-Leffler distribution if its distribution function has the following form*

$$F_\beta(x) = \sum_{k=1}^{\infty} \frac{(-1)^{k-1} x^{k\beta}}{\blacksquare(1 + k\beta)}, \tag{7}$$

where $0 < \beta \leq 1$, $x > 0$, and $F_\beta(x) = 0$ for $x \leq 0$. For $0 < \beta < 1$, the Mittag-Leffler distribution is a heavy-tailed generalization of the exponential, and reduces to the exponential distribution when $\beta = 1$.

A Mittag-Leffler random number can be generated using the most convenient expression proposed by Kozubowski and Rachev [20]:

$$\tau_\beta = -\gamma \ln u \left(\frac{\sin(\beta\pi)}{\tan(\beta\pi v)} - \cos(\beta\pi) \right)^{1/\beta}, \tag{8}$$

where γ is the scale parameter, $u, v \in (0, 1)$ are independent uniform random numbers, and τ_β is a Mittag-Leffler random number.

Example 2 (Pareto Distribution). *A random variable is said to subject to Pareto distribution if its cumulative distribution function has the following expression:*

$$F(x) = \begin{cases} 1 - \left(\frac{b}{x}\right)^a, & x \geq b, \\ 0, & x < b, \end{cases} \tag{9}$$

where $b > 0$ is the scale parameter, $a > 0$ is the shape parameter (Pareto's index of inequality).

Example 3 (Cauchy Distribution). *A random variable is said to subject to Cauchy distribution if its cumulative distribution function has the following expression:*

$$F(x) = \frac{1}{\pi} \arctan \left(\frac{2(x - \mu)}{\sigma} \right) + \frac{1}{2}, \tag{10}$$

where μ is the location parameter, σ is the scale parameter.

Example 4 (Weibull Distribution). *A random variable is said to subject to Weibull distribution if it has a tail function F as follows:*

$$F(x) = e^{-(x/\kappa)^\xi}, \tag{11}$$

where $\kappa > 0$ is the scale parameter, $\xi > 0$ is the shape parameter. If and only if $\xi < 1$, the Weibull distribution is a heavy-tailed distribution.

3.2. Improving CS with Different Heavy-Tailed Probability Distributions

For swarm-based search algorithms, randomness plays a significant role in both exploration and exploitation or diversification and intensification [17]. It is very necessary to discuss the optimal randomness in swarm-based search algorithms. Randomness is normally realized by employing pseudorandom numbers, based on some common stochastic processes. Generally, randomization is achieved by simple random numbers that are drawn from a uniform distribution or a normal distribution. But in other cases, more sophisticated randomization approaches are considered, for example, random walks and Lévy flights. Here, we have to say the foragers should move following a heavy-tailed distribution since Lévy distribution is a simple heavy-tailed distribution which is easy to analyze. There are many other heavy-tailed distributions such as Mittag-Leffler distribution, Pareto distribution, Cauchy distribution, and Weibull distribution, and large steps still frequently happen when using them to generate the steps. In this paper, we mainly focus on the discussion on the impact of different heavy-tailed distributions on the performance of swarm-based search algorithms. In the study, CS is taken as a representative method of swarm-based optimization algorithms, and the results can be generalized to other swarm-based search algorithms.

In this section, four randomness-enhanced cuckoo search algorithms are proposed in this paper. Specifically, the following modified CS methods are considered: (1) CS with the Mittag-Leffler distribution, denoted as CSML; (2) CS with the Pareto distribution, denoted as CSP; (3) CS with the Cauchy distribution, denoted as CSC; (4) CS with the Weibull distribution, referred to CSW. In the modified CS methods, the aforementioned four different heavy-tailed probability distributions are respectively used to be integrated into CS instead of the original Lévy flights in the global explorative random walk. By using these heavy-tailed probability distributions, the updating Equation (1) can be reformulated as follows

$$U_i^t = X_i^t + \alpha \otimes \text{Mittag} - \text{Leffler}(\beta, \gamma) \otimes (X_i^t - X_{best}), \tag{12}$$

$$U_i^t = X_i^t + \alpha \otimes \text{Pareto}(b, a) \otimes (X_i^t - X_{best}), \tag{13}$$

$$U_i^t = X_i^t + \alpha \otimes \text{Cauchy}(\mu, \sigma) \otimes (X_i^t - X_{best}), \tag{14}$$

$$U_i^t = X_i^t + \alpha \otimes \text{Weibull}(\xi, \kappa) \otimes (X_i^t - X_{best}), \tag{15}$$

where $\text{Mittag} - \text{Leffler}(\beta, \gamma)$ in Equation (12) denotes a random number drawn from Mittag-Leffler distribution; $\text{Pareto}(b, a)$ in Equation (13) represents a random number drawn from Cauchy distribution; $\text{Cauchy}(\mu, \sigma)$ in Equation (14) denotes a random number drawn from Cauchy distribution; $\text{Weibull}(\alpha, \kappa)$ in Equation (15) means a random number drawn from Weibull distribution. Compared with the standard CS algorithm, the differences of randomness-enhanced cuckoo search methods lie in line 9 from Algorithm 1.

Remark 1. *In this paper, our emphasis is to study the effects of different heavy-tailed distributions on the swarm-based search algorithms.*

Remark 2. *Since CS is a popular swarm-based search algorithm, we only use it as an representative. Similar analyses for optimal randomness can be applied to other swarm-based algorithms.*

Remark 3. *The source code of randomness-enhanced cuckoo search algorithms (namely CSML, CSP, CSC, CSW), written in Matlab, is available at https://www.mathworks.com/matlabcentral/fileexchange/71758-optimal-randomness-in-swarm-based-search.*

4. Experimental Results

This study focuses on discussing the effectiveness and efficiency of the proposed randomness-enhanced CS algorithms. To fulfill this purpose, extensive experiments are carried out on a test suite of 20 benchmark functions. The superiority of randomness-enhanced CS algorithms over the standard CS is tested, then a scalability study is performed. Finally, an application to parameter identification of fractional-order chaotic systems is also investigated.

4.1. Experimental Setup

For parameter settings of CS, CSML, CSP, CSC and CSW, the probability P_a is set to 0.25 [7], the scaling factor α is set to 0.01. The proposed randomness-enhanced CS algorithms introduce new parameters to CS: the scale parameter γ and the Mittag-Leffler index β in CSML; the scale parameter b and the shape parameter a in CSP; the location parameter μ and the scale parameter σ in CSC; the scale parameter $\kappa > 0$ and the shape parameter ζ in CSW. As for these newly introduced parameters, their values are given in Table 1 after analysis in Section 4.2.

Table 1. Parameters for randomness-enhanced cuckoo search (CS) algorithms.

Distribution	Algorithm	Parameters
Mittag-Leffler distribution	CSML	$\gamma = 4.5, \beta = 0.8$
Pareto distribution	CSP	$a = 1.5, b = 4.5$
Cauchy distribution	CSC	$\sigma = 4.5, \mu = 0.8$
Weibull distribution	CSW	$\zeta = 0.3, \kappa = 4$

All the selected benchmark functions are minimization problems, and a more detailed description of these benchmark functions can be found in References [21,22]. The test suite consists of 20 unconstrained single-objective benchmark functions with different characteristics, including unimodal (F_{sph}, F_{ros}), multimodal (F_{ack}, F_{grw}, F_{ras}, F_{sch}, F_{sal}, F_{wht}, F_{pn1} and F_{pn2}) and rotated and/or shifted functions (F_1-F_{10}) as described in Appendix A. Moreover, the population size satisfies $NP = D$ where D denotes the dimension of the problem unless a change is mentioned.

In the experimental studies, the maximum number of function evaluations (namely Max_FEs) is taken as the termination criterion and set to $10,000 \times D$. The average of the function error value $f(X_{best}) - f(X^*)$ is used to assess the optimization performance, where X_{best} is the best solution found by the algorithms in each run and X^* is the actual global optimal solution of the test function. All the algorithms are evaluated for 50 times and the averaged experimental results are recorded for each benchmark function respectively. Besides, two non-parametric statistical tests for independent samples are taken to detect the differences between the proposed algorithm and the compared algorithms. The tests contain the Wilcoxon signed-rank test at the 5% significance level and the Friedman test. The symbol "‡", "†" and "≈" respectively denote the average performance gained by the chosen approach is weaker than, better than, and similar to the compared algorithm. Meanwhile, the best experimental results for each benchmark problem are marked in boldface, for clarity.

4.2. Parameter Tuning

From Section 3.2, it can be seen that each of the four randomness-enhanced CS algorithms brings two new user-defined parameters, for example, the scale parameter γ and the Mittag-Leffler index β in CSML. To illustrate the impact of these two parameters on the optimization results and to offer reference values to users of our algorithm, parameter analyses are conducted in advance and corresponding experiments are performed on unimodal function F_{sph} and multimodal function F_{ack} with dimension D set to 30. The optimal value of selected benchmark functions is 0. $10,000 \times D$ is the default value for Max_FEs. 15 independent runs are carried out for each parameter setting to reduce statistical sampling effects. The experimental results are plotted in Figure 1. For simplicity of description, only the result of parameter tuning for CSML is shown here, and the same operation is conducted on CSP, CSC, and CSW. In Figure 1a, γ varies within interval [0.5, 4.5] in steps of 0.5, β varies from 0.1 to 0.9 in steps of 0.1, and 'Error' represents the average value of the differences between the benchmark function value and its optimal value over 15 independent runs.

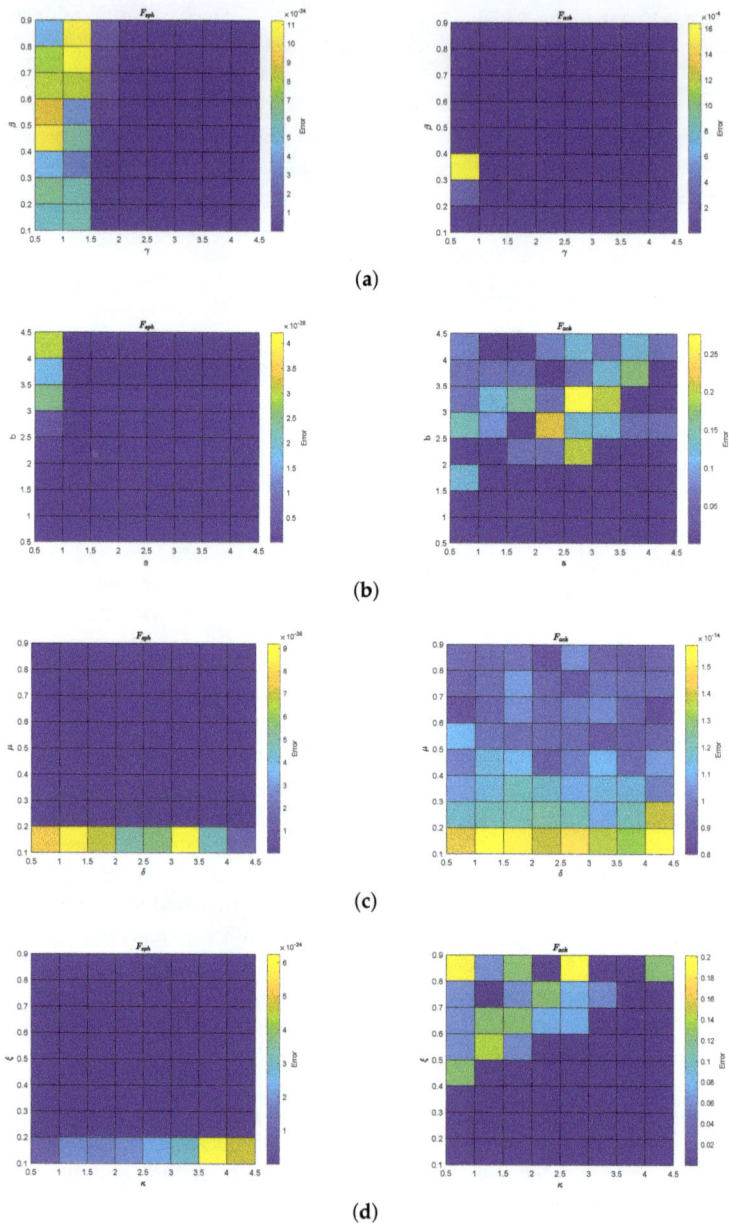

Figure 1. Impact of user-defined parameter values of CSML, CSP, CSC and CSW on the results for selected benchmark functions. (**a**) CS with the Mittag-Leffler distribution, CSML. (**b**) CS with the Pareto distribution, CSP. (**c**) CS with the Cauchy distribution, CSC. (**d**) CS with the Weibull distribution, CSW.

From Figure 1a, we can see that the Mittag-Leffler index β, in general, has a slight effect on the performance of CSML, whereas the value of scale parameter γ shows a more significant impact on the experimental results. According to the right part of each subfigure in Figure 1a, the larger the value of scale parameter γ is, the better the performance of CSML will be. In view of the above considerations, we set the values of γ and β to 4.5 and 0.8 for all the experiments being conducted in the next subsections. For Pareto distribution, Cauchy distribution and Weibull distribution, the same parameter analysis is performed according to Figure 1b–d. The user-defined parameter values for all the randomness-enhanced CS algorithms are listed in Table 1.

4.3. Performance Evaluation of Randomness-Enhanced Cs Algorithms

In this section, lots of experiments are performed in order to probe into the effectiveness and efficiency of different heavy-tailed distributions on the performance of CS, and meanwhile, to decide the optimal randomness in improving CS. In our experiments, the standard CS and four proposed randomness-enhanced CS algorithms (namely, CSML, CSP, CSC, and CSW) are tested on 20 test functions where D is set to 30. The experimental results are presented in Table 2.

Table 2. Comparisons between CS and four randomness-enhanced CS algorithms at $D = 30$.

Fun	CS	CSML	CSP	CSC	CSW
F_{sph}	9.58×10^{-31}	4.90×10^{-54}‡	$\mathbf{4.74\times10^{-59}}$‡	1.17×10^{-57}‡	4.40×10^{-51}‡
F_{ros}	1.20×10^{1}	5.22×10^{0}‡	3.10×10^{0}‡	$\mathbf{2.74\times10^{0}}$‡	8.62×10^{0}‡
F_{ack}	7.70×10^{-13}	1.06×10^{-14}‡	1.07×10^{-14}‡	9.56×10^{-15}‡	$\mathbf{8.28\times10^{-15}}$‡
F_{grw}	7.11×10^{-17}	0.00×10^{0}‡	0.00×10^{0}‡	0.00×10^{0}‡	0.00×10^{0}‡
F_{ras}	2.32×10^{1}	1.38×10^{1}‡	1.88×10^{1}‡	1.49×10^{1}‡	$\mathbf{8.34\times10^{0}}$‡
F_{sch}	1.57×10^{3}	5.37×10^{2}‡	1.32×10^{3}‡	4.80×10^{2}‡	$\mathbf{3.56\times10^{1}}$‡
F_{sal}	3.76×10^{-1}	2.96×10^{-1}‡	3.00×10^{-1}‡	2.84×10^{-1}‡	$\mathbf{2.20\times10^{-1}}$‡
F_{wht}	3.73×10^{2}	2.00×10^{2}‡	2.49×10^{2}‡	2.27×10^{2}‡	$\mathbf{1.93\times10^{2}}$‡
F_{pn1}	2.07×10^{-3}	$\mathbf{1.57\times10^{-32}}$‡	$\mathbf{1.57\times10^{-32}}$‡	2.07×10^{-3}≈	$\mathbf{1.57\times10^{-32}}$‡
F_{pn2}	4.82×10^{-28}	1.35×10^{-32}‡	1.35×10^{-32}‡	1.35×10^{-32}‡	1.35×10^{-32}‡
F_1	6.48×10^{-30}	0.00×10^{0}‡	0.00×10^{0}‡	0.00×10^{0}‡	0.00×10^{0}‡
F_2	1.05×10^{-2}	1.10×10^{-3}‡	2.77×10^{-4}‡	1.40×10^{-3}‡	1.23×10^{-2}†
F_3	$\mathbf{2.17\times10^{6}}$	3.04×10^{6}†	2.99×10^{6}†	3.25×10^{6}†	3.61×10^{6}†
F_4	1.79×10^{3}	4.98×10^{2}‡	$\mathbf{3.58\times10^{2}}$‡	4.02×10^{2}‡	5.51×10^{2}‡
F_5	3.17×10^{3}	2.44×10^{3}‡	1.98×10^{3}‡	2.11×10^{3}‡	$\mathbf{1.94\times10^{3}}$‡
F_6	2.78×10^{1}	1.57×10^{1}‡	$\mathbf{9.91\times10^{0}}$‡	1.23×10^{1}‡	1.59×10^{1}‡
F_7	$\mathbf{1.34\times10^{-3}}$	2.22×10^{-3}†	5.79×10^{-3}†	3.73×10^{-3}†	2.49×10^{-3}†
F_8	$\mathbf{2.09\times10^{1}}$	$\mathbf{2.09\times10^{1}}$≈	$\mathbf{2.09\times10^{1}}$≈	$\mathbf{2.09\times10^{1}}$≈	$\mathbf{2.09\times10^{1}}$≈
F_9	2.84×10^{1}	1.30×10^{1}‡	2.74×10^{1}‡	1.28×10^{1}‡	$\mathbf{6.81\times10^{0}}$‡
F_{10}	1.69×10^{2}	1.21×10^{2}‡	1.31×10^{2}‡	1.18×10^{2}‡	$\mathbf{1.03\times10^{2}}$‡
‡/≈/†	-	17/1/2	17/1/2	16/2/2	16/1/3
p-value	-	8.97×10^{-3}	1.00×10^{-2}	1.00×10^{-2}	1.87×10^{-2}
Avg. rank	4.35	2.78	2.88	2.58	**2.43**

"‡", "†" and "≈" respectively denote the performance of CS is worse than, better than, and similar to those of the proposed algorithms according to the Wilcoxon's rank test at a 0.05 significance level.

According to Table 2, it can be clearly found that CS with different heavy-tailed probability distributions provides significantly better results when compared with the original CS. Specifically speaking, in terms of the total number of "‡/≈ /†", CS is inferior to CSML, CSP, CSC, and CSW on 17, 17, 16 and 16 test functions, similar to CSML, CSP, CSC and CSW on 1, 1, 2 and 1 test functions, and superior to CSML, CSP, CSC, and CSW on 2, 2, 2 and 3 test functions, respectively. It is worth noting that CSML, CSP, CSC and CSW are capable of achieving the global optimum on test problem F_{grw} and F_1, while CS does not. Moreover, all the *p*-values are less than 0.05. These results suggest that CSML, CSP, CSC, and CSW are able to significantly improve the performance of CS for the test functions at $D = 30$. The comprehensive ranking orders are CSW, CSC, CSML, CSP, and CS in a descending manner. This indicates that the integration of different heavy-tailed probability distributions into CS not only retains the merit of CS but also performs even better. Besides, the Weibull distribution performs the best in enhancing the search ability of CS, that is, CSW is supposed to be the optimal randomness in improving CS among all the comparison methods for solving benchmark problems at $D = 30$.

To further discuss the convergence speed of the four randomness-enhanced CS algorithms, several test problems (namely F_{sph}, F_{grw}, F_1 and F_{10}) at $D = 30$ are selected to plot the convergence curves of the averages of the function error values within Max_FEs over 50 independent runs, which are presented in Figure 2. From Figure 2, it can be observed that CSML, CSP, CSC, and CSW converge outstandingly faster than CS according to the convergence curves. In summary, it can be concluded that the standard CS algorithm can be improved by integrating different heavy-tailed probability distributions rather than Lévy distribution into it.

Besides, to analyze the reasons for different performances among the four proposed randomness-enhanced CS algorithms, the jump lengths of CS, CSML, CSP, CSC, and CSW (namely, $\alpha \otimes \text{Lévy}(\lambda)$, $\alpha \otimes \text{MittagLeffler}(\beta, \gamma)$, $\alpha \otimes \text{Pareto}(b, a)$, $\alpha \otimes \text{Cauchy}(\mu, \sigma)$, and $\alpha \otimes \text{Weibull}(\xi, \kappa)$) are depicted in Figure 3, where the parameters are given in Table 1 and the scaling factor is set to 0.01. From Figure 3, it can be observed that (1) Lévy distribution and Cauchy distribution are one-sided distribution where all the random numbers are positive, and the other three distributions are two-sided; (2) large steps frequently take place for all distributions; (3) since the tail of Weibull distribution is the lightest, the extreme large steps (compared with its mean) are less likely to happen.

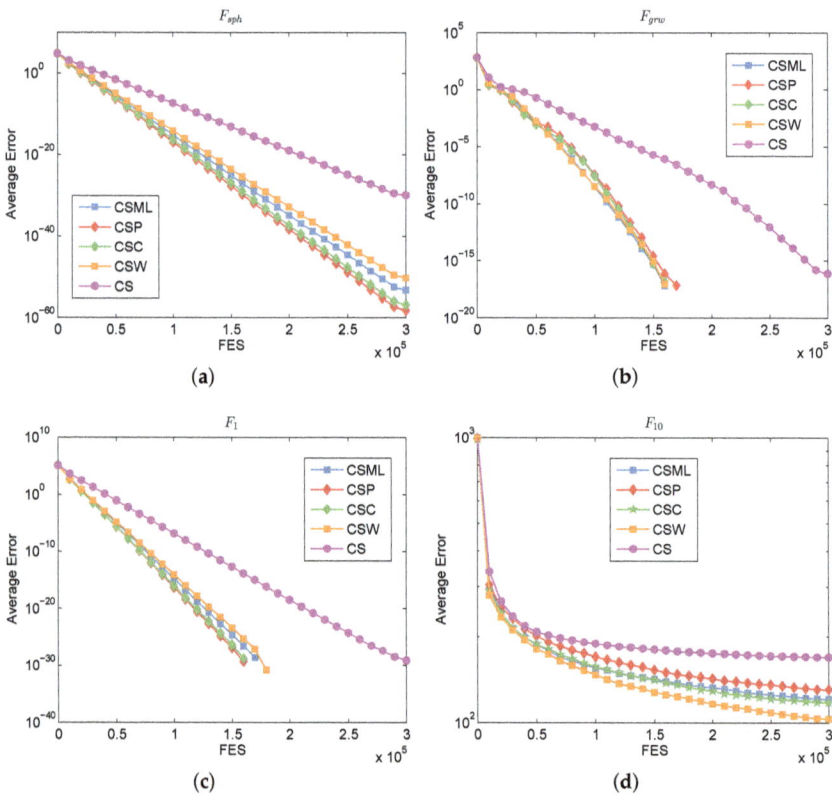

Figure 2. Convergence curves of CS and different improved CS algorithms for selected functions at $D = 30$. (**a**) Sphere's Function F_{sph}. (**b**) Griewank's Function F_{grw}. (**c**) Shifted Sphere Function F_1. (**d**) Shifted Rotated Rastrigin's Function F_{10}.

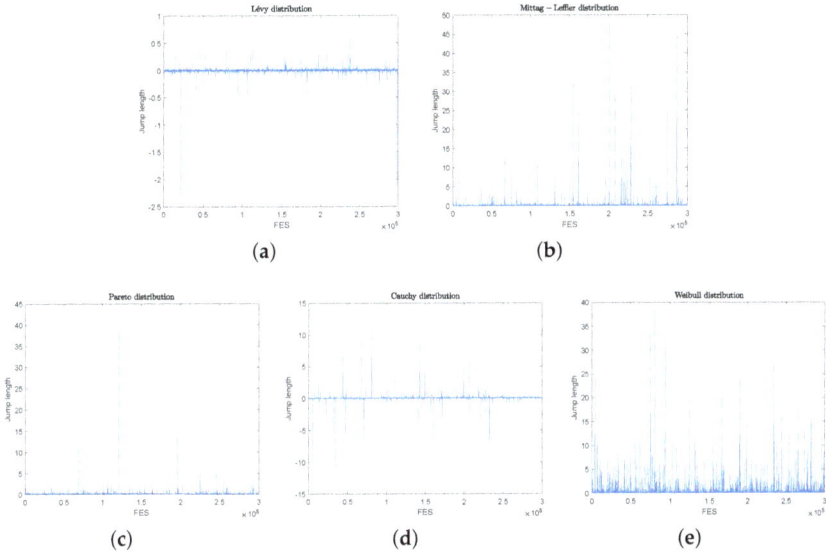

Figure 3. Jump lengths of CS, CSML, CSP, CSC and CSW. (**a**) Lévy distribution of CS. (**b**) Mittag-Leffler distribution of CSML. (**c**) Pareto distribution of CSP. (**d**) Cauchy distribution of CSC. (**e**) Weibull distribution of CSW.

4.4. Scalability Study

In this section, a scalability study comparing with the standard CS algorithm is conducted in order to study the effect of problem size on the performance of the four proposed randomness-enhanced CS algorithms. We carry out experiments on the 20 benchmark functions with dimension D set to 10 and 50. When $D = 10$, the population size is chosen as $NP = 30$; meanwhile, when $D = 50$, the population size is selected as $NP = D$. All the other control parameters are kept unchanged. The experimental results achieved by CS and four proposed randomness-enhanced CS algorithms at $D = 10$ and $D = 50$ are listed in Tables 3 and 4, respectively, and the results of the Wilcoxon signed-rank test are also given in the tables.

According to Table 3, CSML, CSP, CSC, and CSW are significantly better than CS on 7, 17, 18 and 19 test functions, similar to CS on 0, 1, 1 and 1 test functions, and worse than CS on 13, 2, 1 and 0 test functions, respectively. The comprehensive ranking orders in the case of $D = 10$ are CSW, CSC, CSP, CS, and CSML in descending manner. The results show that the performance improvement of using different heavy-tailed probability distributions persists expect CSML when the problem dimension reduces to 10. In the case of $D = 50$, it can be observed from Table 4 that CSML, CSP, CSC and CSW perform better than CS on 16, 14, 16 and 16 test functions, to CS on 1, 1, 1 and 1 test functions, and worse than CS on 3, 5, 3 and 3 test functions, respectively. Meanwhile, the corresponding comprehensive ranking orders when $D = 50$ are CSC, CSML, CSP, CSW and CS. In general, we can draw conclusions that the advantages of four randomness-enhanced CS algorithms over the standard CS are overall stable when the problem dimension increases, except CSML which deteriorates to a certain extent when D set to 10. Furthermore, regarding to the different comprehensive ranking orders obtained at every dimension, it is pointed out that CS with Lévy flights seems not the optimal randomness when compared with those using different heavy-tailed probability distributions in CS.

Table 3. Comparisons between CS and four randomness-enhanced CS algorithms at $D = 10$.

Fun	CS	CSML	CSP	CSC	CSW
F_{sph}	4.87×10^{-26}	$3.39\times10^{-31}\ddagger$	$\mathbf{4.21\times10^{-48}\ddagger}$	$2.04\times10^{-46}\ddagger$	$2.48\times10^{-46}\ddagger$
F_{ros}	9.63×10^{-1}	$3.02\times10^{1}\dagger$	$\mathbf{1.12\times10^{-1}\ddagger}$	$1.75\times10^{-1}\ddagger$	$2.97\times10^{-1}\ddagger$
F_{ack}	4.16×10^{-11}	$2.50\times10^{-14}\ddagger$	$4.37\times10^{-15}\ddagger$	$4.44\times10^{-15}\ddagger$	$4.44\times10^{-15}\ddagger$
F_{grw}	3.44×10^{-2}	$\mathbf{0.00\times10^{0}\ddagger}$	$2.22\times10^{-2}\ddagger$	$2.07\times10^{-2}\ddagger$	$1.44\times10^{-2}\ddagger$
F_{ras}	3.00×10^{0}	$6.93\times10^{1}\dagger$	$2.25\times10^{0}\ddagger$	$2.95\times10^{-1}\ddagger$	$\mathbf{2.26\times10^{-9}\ddagger}$
F_{sch}	6.72×10^{1}	$3.80\times10^{3}\dagger$	$1.38\times10^{1}\ddagger$	$6.91\times10^{-3}\ddagger$	$\mathbf{1.27\times10^{-4}\ddagger}$
F_{sal}	1.04×10^{-1}	$4.78\times10^{-1}\dagger$	$\mathbf{9.99\times10^{-2}\ddagger}$	$9.99\times10^{-2}\ddagger$	$9.99\times10^{-2}\ddagger$
F_{wht}	2.40×10^{1}	$9.89\times10^{2}\dagger$	$1.54\times10^{1}\ddagger$	$1.01\times10^{1}\ddagger$	$\mathbf{5.72\times10^{0}\ddagger}$
F_{pn1}	1.96×10^{-16}	$2.01\times10^{-28}\ddagger$	$4.71\times10^{-32}\ddagger$	$4.71\times10^{-32}\ddagger$	$4.71\times10^{-32}\ddagger$
F_{pn2}	4.86×10^{-23}	$9.17\times10^{-30}\ddagger$	$1.35\times10^{-32}\ddagger$	$1.35\times10^{-32}\ddagger$	$1.35\times10^{-32}\ddagger$
F_{1}	4.13×10^{-26}	$\mathbf{0.00\times10^{0}\ddagger}$	$\mathbf{0.00\times10^{0}\ddagger}$	$\mathbf{0.00\times10^{0}\ddagger}$	$\mathbf{0.00\times10^{0}\ddagger}$
F_{2}	8.16×10^{-14}	$3.73\times10^{2}\dagger$	$1.33\times10^{-21}\ddagger$	$4.54\times10^{-19}\ddagger$	$\mathbf{1.51\times10^{-16}\ddagger}$
F_{3}	$\mathbf{2.08\times10^{2}}$	$1.70\times10^{7}\dagger$	$7.20\times10^{2}\dagger$	$6.78\times10^{2}\dagger$	$8.31\times10^{2}\dagger$
F_{4}	1.01×10^{-5}	$1.96\times10^{4}\dagger$	$\mathbf{1.46\times10^{-9}\ddagger}$	$1.20\times10^{-8}\ddagger$	$4.82\times10^{-8}\ddagger$
F_{5}	9.30×10^{-5}	$6.82\times10^{3}\dagger$	$\mathbf{6.13\times10^{-10}\ddagger}$	$5.11\times10^{-9}\ddagger$	$9.27\times10^{-9}\ddagger$
F_{6}	9.78×10^{-1}	$4.11\times10^{1}\dagger$	$6.38\times10^{-1}\ddagger$	$3.48\times10^{-1}\ddagger$	$\mathbf{2.69\times10^{-1}\ddagger}$
F_{7}	5.33×10^{-2}	$\mathbf{1.07\times10^{-3}\ddagger}$	$5.91\times10^{-2}\ddagger$	$4.72\times10^{-2}\ddagger$	$4.39\times10^{-2}\ddagger$
F_{8}	2.04×10^{1}	$2.11\times10^{1}\dagger$	$2.04\times10^{1}\approx$	$2.04\times10^{1}\approx$	$\mathbf{2.03\times10^{1}\ddagger}$
F_{9}	2.75×10^{0}	$7.37\times10^{1}\dagger$	$1.80\times10^{0}\ddagger$	$1.79\times10^{-1}\ddagger$	$\mathbf{2.35\times10^{-10}\ddagger}$
F_{10}	1.99×10^{1}	$2.89\times10^{2}\dagger$	$1.63\times10^{1}\ddagger$	$1.59\times10^{1}\ddagger$	$\mathbf{1.43\times10^{1}\ddagger}$
$\ddagger/\approx/\dagger$	-	7/0/13	17/1/2	18/1/1	19/1/0
Avg. rank	4.10	4.28	2.28	2.25	**2.10**

"$\ddagger$", "$\dagger$" and "$\approx$" respectively denote the performance of CS is worse than, better than, and similar to those of the proposed algorithms according to the Wilcoxon's rank test at a 0.05 significance level.

Table 4. Comparison results between CS and four randomness-enhanced CS algorithms at $D = 50$.

Fun	CS	CSML	CSP	CSC	CSW
F_{sph}	3.79×10^{-17}	$3.47\times10^{-31}\ddagger$	$\mathbf{7.41\times10^{-36}\ddagger}$	$5.75\times10^{-32}\ddagger$	$2.73\times10^{-24}\ddagger$
F_{ros}	4.22×10^{1}	$3.07\times10^{1}\ddagger$	$\mathbf{2.82\times10^{1}\ddagger}$	$2.99\times10^{1}\ddagger$	$3.41\times10^{1}\ddagger$
F_{ack}	2.85×10^{-2}	$2.43\times10^{-14}\ddagger$	$2.05\times10^{-14}\ddagger$	$2.05\times10^{-14}\ddagger$	$\mathbf{7.40\times10^{-13}\ddagger}$
F_{grw}	1.93×10^{-10}	$\mathbf{0.00\times10^{0}\ddagger}$	$\mathbf{0.00\times10^{0}\ddagger}$	$\mathbf{0.00\times10^{0}\ddagger}$	$\mathbf{0.00\times10^{0}\ddagger}$
F_{ras}	8.44×10^{1}	$\mathbf{6.80\times10^{1}\ddagger}$	$8.69\times10^{1}\dagger$	$7.54\times10^{1}\ddagger$	$7.37\times10^{1}\ddagger$
F_{sch}	4.87×10^{3}	$4.14\times10^{3}\ddagger$	$6.05\times10^{3}\dagger$	$4.38\times10^{3}\ddagger$	$\mathbf{2.38\times10^{3}\ddagger}$
F_{sal}	6.69×10^{-1}	$4.68\times10^{-1}\ddagger$	$4.87\times10^{-1}\ddagger$	$4.22\times10^{-1}\ddagger$	$\mathbf{3.68\times10^{-1}\ddagger}$
F_{wht}	1.36×10^{3}	$\mathbf{9.58\times10^{2}\ddagger}$	$1.21\times10^{3}\ddagger$	$1.09\times10^{3}\ddagger$	$1.13\times10^{3}\ddagger$
F_{pn1}	8.13×10^{-3}	$6.74\times10^{-28}\ddagger$	$1.04\times10^{-27}\ddagger$	$\mathbf{7.21\times10^{-30}\ddagger}$	$1.47\times10^{-23}\ddagger$
F_{pn2}	3.25×10^{-14}	$1.02\times10^{-29}\ddagger$	$\mathbf{1.57\times10^{-32}\ddagger}$	$1.44\times10^{-30}\ddagger$	$2.01\times10^{-23}\ddagger$
F_{1}	1.40×10^{-16}	$\mathbf{0.00\times10^{0}\ddagger}$	$\mathbf{0.00\times10^{0}\ddagger}$	$\mathbf{0.00\times10^{0}\ddagger}$	$3.57\times10^{-24}\ddagger$
F_{2}	2.34×10^{2}	$3.57\times10^{2}\dagger$	$\mathbf{1.86\times10^{2}\ddagger}$	$4.49\times10^{2}\dagger$	$8.59\times10^{2}\dagger$
F_{3}	$\mathbf{8.53\times10^{6}}$	$1.66\times10^{7}\dagger$	$1.47\times10^{7}\dagger$	$1.83\times10^{7}\dagger$	$1.85\times10^{7}\dagger$
F_{4}	2.72×10^{4}	$1.99\times10^{4}\ddagger$	$\mathbf{1.72\times10^{4}\ddagger}$	$1.91\times10^{4}\ddagger$	$1.88\times10^{4}\ddagger$
F_{5}	1.06×10^{4}	$6.95\times10^{3}\ddagger$	$6.49\times10^{3}\ddagger$	$6.65\times10^{3}\ddagger$	$6.30\times10^{+03}\ddagger$
F_{6}	6.38×10^{1}	$3.90\times10^{1}\ddagger$	$4.15\times10^{1}\ddagger$	$\mathbf{3.63\times10^{1}\ddagger}$	$4.43\times10^{1}\ddagger$
F_{7}	1.30×10^{-3}	$\mathbf{1.81\times10^{-3}\dagger}$	$3.56\times10^{-3}\dagger$	$2.43\times10^{-3}\dagger$	$3.64\times10^{-3}\dagger$
F_{8}	2.11×10^{1}	$2.11\times10^{1}\approx$	$2.11\times10^{1}\approx$	$2.11\times10^{1}\approx$	$2.11\times10^{1}\approx$
F_{9}	1.24×10^{2}	$7.04\times10^{1}\ddagger$	$1.27\times10^{2}\dagger$	$7.47\times10^{1}\ddagger$	$\mathbf{6.50\times10^{1}\ddagger}$
F_{10}	3.87×10^{2}	$2.87\times10^{2}\ddagger$	$3.13\times10^{2}\ddagger$	$2.85\times10^{2}\ddagger$	$\mathbf{2.69\times10^{2}\ddagger}$
$\ddagger/\approx/\dagger$	-	16/1/3	14/1/5	16/1/3	16/1/3
Avg. rank	4.10	2.58	2.70	**2.55**	3.08

"$\ddagger$", "$\dagger$" and "$\approx$" respectively denote the performance of CS is worse than, better than, and similar to those of the proposed algorithms according to the Wilcoxon's rank test at a 0.05 significance level.

4.5. Application to Parameter Identification of Fractional-Order Chaotic Systems

In this section, the four proposed randomness-enhanced CS algorithms (namely, CSML, CSP, CSC and CSW) are applied to identify unknown parameters of fractional-order chaotic systems, which is a critical issue in chaos control and synchronization. Our main task of this section is to further demonstrate that improving CS with different heavy-tailed probability distributions can also effectively tackle the real-world complex optimization problems besides the benchmark problems. In fact, by using a non-Lyapunov way according to problem formulation suggested in Reference [23], the nonlinear function optimization can be converted to from parameter identification of uncertain fractional-order chaotic systems.

In the numerical simulation, the fractional-order financial system [24] under the Caputo definition is taken for example, which can be described as

$$\begin{cases} {}_0D_t^{q_1}x(t) = z(t) + x(t)(y(t) - a), \\ {}_0D_t^{q_2}y(t) = 1 - by(t) - x^2(t), \\ {}_0D_t^{q_3}z(t) = -x(t) - cz(t), \end{cases} \tag{16}$$

where q_1, q_2, q_3 and a, b, c are fractional orders and systematic parameters. When $(q_1, q_2, q_3) = (1, 0.95, 0.99)$, $(a, b, c) = (1, 0.1, 1)$ and initial point $(x_0, y_0, z_0) = (2, -1, 1)$, the system above is chaotic.

Suppose the structure of system (16) is known and the systematic parameters a, b, c are unknown, then the objective function is defined as

$$F = \sum_{k=1}^{N} \|X_k - \hat{X}_k\|^2, \tag{17}$$

where X_k and $\hat{X}_k$ denote the state vector of system (16) and its estimated system at time k, respectively. $k = 1, 2, \ldots, M$ is the sampling time point and M denotes the length of data used for parameter estimation. $\|\cdot\|$ denotes Euclid norm. Parameter identification can be achieved by searching suitable (a^*, b^*, c^*) such that the objective function (17) is minimized, that is,

$$(a^*, b^*, c^*) = \arg \min_{(\hat{a}, \hat{b}, \hat{c}) \in \blacksquare} F, \tag{18}$$

where $\blacksquare$ is the searching space admitted for systematic parameters. Figure 4 depicts the distribution figure of system (16) for the objective function values.

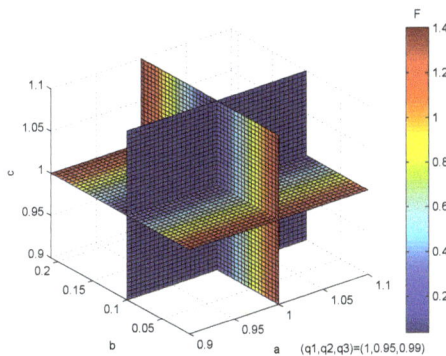

Figure 4. Distribution of the objective function values for system (16).

The validation of the proposed methods in this paper is further proved by comparing CSML, CSP, CSC and CSW with the standard CS algorithm for parameter identification. In the simulations, the maximum iteration number is set to 200 and the population size is set to 40. For the system to be identified, the step size is set to 0.005 and the number of samples set to 200. In addition, it is worth mentioning that the same computation effort is used in implementation for all the compared algorithms to make a fair comparison. Table 5 lists the statistical results of the average identified values, the corresponding relative error values and the objective function values for system (16). From Table 5, it can be clearly observed that all the four proposed randomness-enhanced CS algorithms outperform CS according to the average objective function values and they are able to generate estimated values with much higher accuracy than CS. Besides, it can be seen that CSP surpasses CS, CSML, CSW and CSC in obtaining the best average identified values, the corresponding relative error values and the objective function values.

Table 5. Statistical results of different methods for system (16), in terms of the average estimated values, the relative error values and objective function values.

Method	CS	CSML	CSP	CSC	CSW
a	0.999999825481796	0.999999979386471	**1.000000001165006**	0.999999930875086	0.999999994619958
$\frac{\|a-1.00\|}{1.00}$	1.75×10^{-7}	2.28×10^{-8}	**1.17×10^{-9}**	6.91×10^{-8}	5.38×10^{-9}
b	0.100000078306700	0.100000006492360	**0.099999999732393**	0.100000038684769	0.100000001325757
$\frac{\|b-0.10\|}{0.10}$	7.83×10^{-7}	1.12×10^{-7}	**2.68×10^{-9}**	3.87×10^{-7}	2.06×10^{-8}
c	1.000000126069434	0.999999979588057	**0.999999995606294**	0.999999876500337	0.999999979353103
$\frac{\|c-1.00\|}{1.00}$	1.26×10^{-7}	4.61×10^{-8}	**4.39×10^{-9}**	1.23×10^{-7}	1.33×10^{-8}
$F_{Avg\pm Std}$	1.07×10^{-5}	4.75×10^{-7}	**7.46×10^{-8}**	1.89×10^{-6}	1.03×10^{-7}
F_{Std}	5.46×10^{-6}	2.74×10^{-7}	**3.29×10^{-8}**	9.38×10^{-7}	6.12×10^{-8}

a, b and c are parameters to be identified. Their actual values are 1.00, 0.10 and 1.00, respectively.

Moreover, Figure 5 shows the convergence curves of the relative error values of the estimated parameters and objective function values for the corresponding system via CSML, CSP, CSC, CSW and CS. From Figure 5a–c, the relative error values of the estimated values generated by the randomness-enhanced CS algorithms converge to zero more quickly than the original CS. This indicates that CS algorithms with the four different heavy-tailed probability distributions are able to obtain more accurate values of the estimated parameters. In terms of Figure 5d, the objective function values of CSML, CSP, CSC, CSW also decline faster than CS and among which CSP performs the best. It is noteworthy that CSW has a similar convergence curve of objective function values with CSP and can converge to the nearby area of CSP. Therefore, CSW can still be considered as an efficient tool for solving optimization problems.

According to the foregoing discussion, it can be summarized that the randomness-enhanced CS algorithms are able to exactly identify the unknown specific parameters of the fractional-order system (16) with better effectiveness and robustness and CSP together with CSW may be treated as a useful tool for handling the problem of parameter identification.

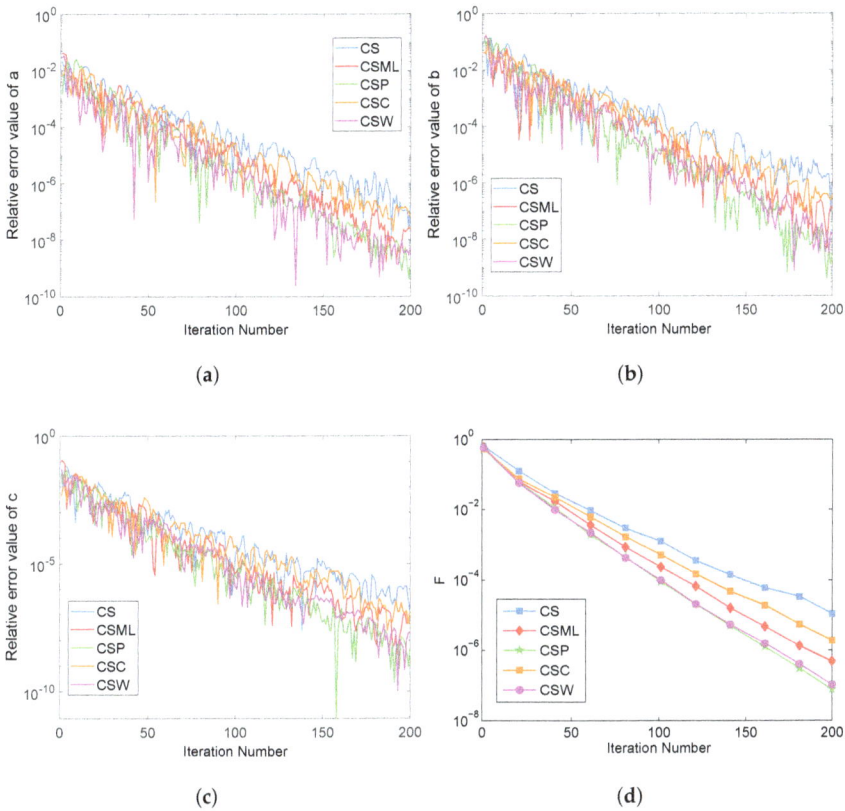

Figure 5. The convergence curves of the relative error values and objective function values for system (16). (**a**) The relative error value of *a*. (**b**) The relative error value of *b*. (**c**) The relative error value of *c*. (**d**) Objective function value of *F*.

5. Conclusions

The purpose of this paper is to study the optimal randomness in swarm-based search algorithms. In the study, CS is taken as a representative method of swarm-based optimization algorithms and the results can be generalized to other swarm-based search algorithms. The impact of different heavy-tailed distributions on the performance of CS is investigated. By replacing Lévy flights with steps generated from other heavy-tailed distributions in CS, four different randomness-enhanced CS algorithms (namely CSML, CSP, CSC and CSW) are presented by applying Mittag-Leffler distribution, Pareto distribution, Cauchy distribution and Weibull distribution, in order to improve the optimization performance of CS. The improvement in effectiveness and efficiency is validated through dedicated experiments. The experimental results indicate that all four proposed randomness-enhanced CS algorithms show a significant improvement in effectiveness and efficiency over the standard CS algorithm. Furthermore, the randomness-enhanced CS algorithms are successfully applied to system identification. In summary, CS with different heavy-tailed probability distributions can be regarded as an efficient and promising tool for solving the real-world complex optimization problems besides the benchmark problems.

Future promising topics can be directed to (1) theoretically analyze the global convergence of randomness-enhanced CS algorithms; (2) do a similar analyses to other swarm-based search algorithms for the optimal randomness; (3) since the search range is always finite for swarm-based search algorithms, it is necessary to study the optimal randomness in a finite range.

Author Contributions: Methodology, J.W. and Y.C. (Yuquan Chen); Conceptualization, Y.C. (YangQuan Chen); Implementation of numerical schemes and Writing of the manuscript was completed by J.W.; Review & Editing by Y.C. (YangQuan Chen); Supervision, Y.C. (YangQuan Chen) and Y.Y. All authors read and approved the final manuscript.

Funding: This work is supported by the Fundamental Research Funds for the Central Universities (No. 2017YJS200), China Scholarship Council (No. 201807090092) and the National Nature Science Foundation of China (No. 61772063).

Acknowledgments: The authors are thankful to the anonymous referees for their invaluable suggestions.

Conflicts of Interest: The authors declare no conflict of interest.

Appendix A

The description of the 20 benchmark functions is given as follows. The formulae of the corresponding benchmark functions are presented in Table A1.

1. F_{sph}: Sphere's Function.
2. F_{ros}: Rosenbrock's Function.
3. F_{ack}: Ackley's Function.
4. F_{grw}: Griewank's Function.
5. F_{ras}: Rastrigin's Function.
6. F_{sch}: Generalized Schwefel's Problem 2.26.
7. F_{sal}: Salomon's Function.
8. F_{wht}: Whitely's Function.
9. F_{pn1}: Generalized Penalized Function 1.
10. F_{pn2}: Generalized Penalized Function 2.
11. F_1: Shifted Sphere Function.
12. F_2: Shifted Schwefel's Problem 1.2.
13. F_3: Shifted Rotated High Conditioned Elliptic Function.
14. F_4: Shifted Schwefel's Problem 1.2 with Noise in Fitness.
15. F_5: Schwefel's Problem 2.6 with global Optimum on Bounds.
16. F_6: Shifted Rosenbrock's Function.
17. F_7: Shifted Rotated Griewank's Function without Bounds.
18. F_8: Shifted Rotated Ackley's Function with Global Optimum on Bounds.
19. F_9: Shifted Rastrigin's Function.
20. F_{10}: Shifted Rotated Rastrigin's Function.

Table A1. Description of the benchmark functions at dimension D used in experiments.

No.	Formula	Range	Optimum
1	$F_{sph}(x) = \sum_{i=1}^{D} x_i^2$	$[-100, 100]^D$	0
2	$F_{ros}(x) = \sum_{i=1}^{D-1}(100(x_{i+1} - x_i^2)^2 + (x_i - 1)^2)$	$[-100, 100]^D$	0
3	$F_{ack}(x) = 20 + \exp(1) - 20\exp\left(-0.2\sqrt{\frac{1}{D}\sum_{i=1}^{D} x_i^2}\right) - \exp\left(\frac{1}{D}\sum_{i=1}^{D}\cos(2\pi x_i)\right)$	$[-32, 32]^D$	0
4	$F_{grw}(x) = \sum_{i=1}^{D}\frac{x_i^2}{4000} - \prod_{i=1}^{D}\cos\frac{x_i}{\sqrt{i}} + 1$	$[-600, 600]^D$	0
5	$F_{ras}(x) = 10D + \sum_{i=1}^{D}(x_i^2 - 10\cos(2\pi x_i))$	$[-5, 5]^D$	0
6	$F_{sch}(x) = 418.9829D - \sum_{i=1}^{D}\left(x_i\sin\left(\sqrt{|x_i|}\right)\right)$	$[-500, 500]^D$	0
7	$F_{sal}(x) = -\cos\left(2\pi\sqrt{\sum_{i=1}^{D}x_i^2}\right) + 0.1\sqrt{\sum_{i=1}^{D}x_i^2} + 1$	$[-100, 100]^D$	0
8	$F_{wht}(x) = \sum_{j=1}^{D}\sum_{i=1}^{D}\left(\frac{y_{i,j}^2}{4000} - \cos(y_{i,j}) + 1\right)$, where $y_{i,j} = 100(x_j - x_i^2)^2 + (1 - x_i)^2$	$[-100, 100]^D$	0
9	$F_{pn1}(x) = \frac{\pi}{D}\left\{10\sin^2(\pi y_1) + \sum_{i=1}^{D-1}(y_i - 1)^2[1 + 10\sin^2(\pi y_{i+1})] + (y_D - 1)^2\right\} + \sum_{i=1}^{D}u(x_i, 10, 100, 4)$, where $y_i = 1 + \frac{1}{4}(x_i + 1)$ and $u(x_i, a, k, m) = \begin{cases} k(x_i - a)^m, & x_i > a, \\ 0, & -a \le x_i \le a, \\ k(-x_i - a)^m, & x_i < -a. \end{cases}$	$[-50, 50]^D$	0
10	$F_{pn2}(x) = 0.1\left\{\sin^2(3\pi x_1) + \sum_{i=1}^{D-1}(x_i - 1)^2[1 + \sin^2(3\pi x_{i+1})] + (x_D - 1)^2[1 + \sin^2(2\pi x_D)]\right\} + \sum_{i=1}^{D}u(x_i, 5, 100, 4)$	$[-50, 50]^D$	0
11	$F_1(x) = \sum_{i=1}^{D}z_i^2 + f_{bias_1}$, where $z = x - o$, o: the shifted global optimum, $f_{bias_1} = -450$	$[-100, 100]^D$	0
12	$F_2(x) = \sum_{i=1}^{D}\left(\sum_{j=1}^{i}z_j\right)^2 + f_{bias_2}$, where $z = x - o$, o: the shifted global optimum, $f_{bias_2} = -450$	$[-100, 100]^D$	0
13	$F_3(x) = \sum_{i=1}^{D}(10^6)^{\frac{i-1}{D-1}}z_i^2 + f_{bias_3}$, where $z = (x - o)*M$, o: the shifted global optimum, M: orthogonal matrix, $f_{bias_3} = -450$	$[-100, 100]^D$	0
14	$F_4(x) = \left(\sum_{i=1}^{D}\left(\sum_{j=1}^{i}z_j\right)^2\right)(1 + 0.4|N(0,1)|) + f_{bias_4}$, where $z = x - o$, o: the shifted global optimum, $f_{bias_4} = -450$	$[-100, 100]^D$	0
15	$F_5(x) = \max\{|A_i x - B_i|\} + f_{bias_5}$, where A is a $D*D$ matrix, $a_{i,j}$ are integer random numbers in the range $[-500, 500]$, $\det(A) \neq 0$, A_i is the ith row of A, $i = 1, ..., D$, $B_i = A_i * o$ is a $D*D$ matrix, o_i are random number in the range $[-100, 100]$ and $f_{bias_5} = -310$	$[-100, 100]^D$	0
16	$F_6(x) = \sum_{i=1}^{D-1}(100(z_i^2 - z_{i+1})^2 + (z_i - 1)^2) + f_{bias_6}$, where $z = x - o + 1$, o: the shifted global optimum, $f_{bias_6} = 390$	$[-100, 100]^D$	0
17	$F_7(x) = \sum_{i=1}^{D}\frac{z_i^2}{4000} - \prod_{i=1}^{D}\cos\left(\frac{z_i}{\sqrt{i}}\right) + 1 + f_{bias_7}$, where $z = (x - o)*M$, $M = M'(1 + 0.3|N(0,1)|)$, M: linear transformation matrix, condition number=3, $f_{bias_7} = -180$ // initial population in $[0, 600]^D$ and no bounds for x	-	0
18	$F_8(x) = -20\exp\left(-0.2\sqrt{\frac{1}{D}\sum_{i=1}^{D}z_i^2}\right) - \exp\left(\frac{1}{D}\sum_{i=1}^{D}\cos(2\pi z_i)\right) + 20 + \exp(1) + f_{bias_8}$, where $z = (x - o)*M$, o: the shifted global optimum, M: linear transformation matrix, condition number=100, $f_{bias_8} = -140$	$[-32, 32]^D$	0
19	$F_9(x) = \sum_{i=1}^{D}(z_i^2 - 10\cos(2\pi z_i) + 10) + f_{bias_9}$, where $z = x - o$, o: the shifted global optimum, $f_{bias_9} = -330$	$[-5, 5]^D$	0
20	$F_{10}(x) = \sum_{i=1}^{D}(z_i^2 - 10\cos(2\pi z_i) + 10) + f_{bias_{10}}$, where $z = (x - o)*M$, M: linear transformation matrix, condition number=2, $f_{bias_{10}} = -330$,	$[-5, 5]^D$	0

References

1. Yang, X.S. *Nature-Inspired Metaheuristic Algorithms*; Luniver Press: Beckington, UK, 2010.
2. Anandakumar, H.; Umamaheswari, K. A bio-inspired swarm intelligence technique for social aware cognitive radio handovers. *Comput. Electr. Eng.* **2018**, *71*, 925–937. [CrossRef]
3. Brezočnik, L.; Fister, I.; Podgorelec, V. Swarm intelligence algorithms for feature selection: A review. *Appl. Sci.* **2018**, *8*, 1521. [CrossRef]
4. Zhao, X.; Wang, C.; Su, J.; Wang, J. Research and application based on the swarm intelligence algorithm and artificial intelligence for wind farm decision system. *Renew. Energy* **2019**, *134*, 681–697. [CrossRef]
5. Dulebenets, M.A. A novel memetic algorithm with a deterministic parameter control for efficient berth scheduling at marine container terminals. *Marit. Bus. Rev.* **2017**, *2*, 302–330. [CrossRef]
6. Karaboga, D.; Basturk, B. A powerful and efficient algorithm for numerical function optimization: Artificial bee colony (ABC) algorithm. *J. Glob. Optim.* **2007**, *39*, 459–471. [CrossRef]
7. Yang, X.S.; Deb, S. Cuckoo search via Lévy flights. In Proceedings of the Nature & Biologically Inspired Computing, Coimbatore, India, 9–11 December 2009; IEEE: Piscataway, NY, USA, 2009; pp. 210–214.
8. Yang, X.S. Firefly algorithms for multimodal optimization. In Proceedings of the International Symposium on Stochastic Algorithms, Sapporo, Japan, 26–28 October 2009; Springer: Berlin/Heidelberg, Germany, 2009; pp. 169–178.
9. Kennedy, J.; Eberhart, R. Particle swarm optimization (PSO). In Proceedings of the IEEE International Conference on Neural Networks, Perth, Australia, 27 November–1 December 1995; pp. 1942–1948.
10. Zheng, H.; Zhou, Y. A novel cuckoo search optimization algorithm based on Gauss distribution. *J. Comput. Inf. Syst.* **2012**, *8*, 4193–4200.
11. Wang, L.; Zhong, Y.; Yin, Y. Nearest neighbour cuckoo search algorithm with probabilistic mutation. *Appl. Soft Comput.* **2016**, *49*, 498–509. [CrossRef]
12. Rakhshani, H.; Rahati, A. Snap-drift cuckoo search: A novel cuckoo search optimization algorithm. *Appl. Soft Comput.* **2017**, *52*, 771–794. [CrossRef]
13. Cui, Z.; Sun, B.; Wang, G.; Xue, Y.; Chen, J. A novel oriented cuckoo search algorithm to improve DV-Hop performance for cyber–physical systems. *J. Parallel Distrib. Comput.* **2017**, *103*, 42–52. [CrossRef]
14. Salgotra, R.; Singh, U.; Saha, S. New cuckoo search algorithms with enhanced exploration and exploitation properties. *Expert Syst. Appl.* **2018**, *95*, 384–420. [CrossRef]
15. Richer, T.J.; Blackwell, T.M. The Lévy particle swarm. In Proceedings of the 2006 IEEE International Conference on Evolutionary Computation, Vancouver, BC, Canada, 16–21 July 2006; IEEE: Piscataway, NY, USA, 2006; pp. 808–815.
16. Pavlyukevich, I. Lévy flights, non-local search and simulated annealing. *J. Comput. Phys.* **2007**, *226*, 1830–1844. [CrossRef]
17. Yang, X.S. *Nature-Inspired Optimization Algorithms*; Elsevier: Amsterdam, The Netherlands, 2014.
18. Yang, X.S.; Deb, S. Engineering optimisation by cuckoo search. *Int. J. Math. Model. Numer. Optim.* **2010**, *1*, 330–343. [CrossRef]
19. Foss, S.; Korshunov, D.; Zachary, S. *An Introduction to Heavy-Tailed and Subexponential Distributions*; Springer: Berlin/Heidelberg, Germany, 2011; Volume 6.
20. Kozubowski, T.J.; Rachev, S.T. Univariate geometric stable laws. *J. Comput. Anal. Appl.* **1999**, *1*, 177–217.
21. Noman, N.; Iba, H. Accelerating differential evolution using an adaptive local search. *IEEE Trans. Evol. Comput.* **2008**, *12*, 107–125. [CrossRef]
22. Suganthan, P.N.; Hansen, N.; Liang, J.J.; Deb, K.; Chen, Y.P.; Auger, A.; Tiwari, S. Problem definitions and evaluation criteria for the CEC 2005 special session on real-parameter optimization. *KanGAL Rep.* **2005**, *2005005*, 2005.

23. Gao, F.; Fei, F.X.; Lee, X.J.; Tong, H.Q.; Deng, Y.F.; Zhao, H.L. Inversion mechanism with functional extrema model for identification incommensurate and hyper fractional chaos via differential evolution. *Expert Syst. Appl.* **2014**, *41*, 1915–1927. [CrossRef]

24. Chen, W.C. Nonlinear dynamics and chaos in a fractional-order financial system. *Chaos Solitons Fractals* **2008**, *36*, 1305–1314. [CrossRef]

MDPI

St. Alban-Anlage 66

4052 Basel

Switzerland

Tel. +41 61 683 77 34

Fax +41 61 302 89 18

www.mdpi.com

Mathematics Editorial Office

E-mail: mathematics@mdpi.com

www.mdpi.com/journal/mathematics